SAFARI NJEMA

A MEMOIR

SAFARI NJEMA

My Journey from Kilimanjaro to America

Lioba M. Moshi

LUMINARE PRESS
WWW.LUMINAREPRESS.COM

Safari Njema: My Journey from Kilimanjaro to America
Copyright © 2024 by Lioba M. Moshi

All rights reserved. This book or any portion thereof may not be reproduced or used in any manner whatsoever without the express written permission of the publisher, except for the use of brief quotations in a book review.

Printed in the United States of America

Luminare Press
442 Charnelton St.
Eugene, OR 97401
www.luminarepress.com

LCCN: 2023909618
ISBN: 979-8-88679-040-5

*For my family in Tanzania and the
United States of America with a special callout to
the first-generation Americans in my family:
Christopher, Monica, Mwayi, Isaiah, and their
offspring as well as the future children of my nephew
Charles and their children. I want them to remember
that it is a right and privilege to be citizens of the
world and to always do good for humanity.*

and

*In memory of my parents
Martha Nzarenau Sambonanga (1925-1995)
John Kifaranja Semali ((1921-2007)*

TABLE OF CONTENTS

Foreword ix
Introduction xi

The Dirt Country Road from Home 1
The Chosen Few. 32
Life in Boarding School 50
Just Being Kids: Surviving Middle School. 79
New Beginnings 103
College Life 137
My First Job. 155
New Opportunities 188

PART 2

OPENING THE DOORS TO SUCCESS

Game Changer 221
The Second Trip to America. 241
Surviving Graduate School in California 255
Where There is a Will, there is a Way 267
An Amazing Tour of Three African Countries 290
Leaving California for Georgia. 333

Up the Ladder at the University of Georgia 350

The African Studies Institute at the
University of Georgia 382

Growing the African Studies Program 399

Shared Experience with My Family 425

Epilogue 463

Acknowledgements 476

Foreword

In my lifetime I have known only a couple of individuals whose work has impacted multiple generations: the late Eugene Odum, the ecologist who developed the ecosystem concept; and Lioba Moshi, the scholar, teacher, and humanitarian who has implemented in the state of Georgia and in her Tanzania homeland the value of cross-cultural education.

Dr. Moshi has taken countless groups of students, teachers, and law enforcement officers from Georgia to Tanzania to exchange ideas with their counterparts there. And she has brought ambassadors and Foreign Service personnel, as well as talented artists and writers, from Tanzania and other African nations to Georgia. She founded the African Studies Center at the University of Georgia and built it into the most popular foreign studies unit on campus.

Most important is her latest project. In her retirement, she has raised funds to build and sustain a home for abandoned and orphaned children in her hometown of Moshi, Tanzania.

I like the photograph of Dr. Moshi on the cover of this book. To me, it epitomizes Lioba's extraordinary life's journey from a village in East Africa to a doctorate from the University of California at Los Angeles to an academic career marked by honors, including the title University Professor of Comparative Literature, into a retirement dedicated to education for Tanzanian children who otherwise would have little opportunity to know the world. Lioba is respected and loved by multitudes.

Betty Jean Craige

Introduction

I often ask myself the following questions:

1. *Why am I here when the friends I went to school with from the first grade to college are gone?*
2. *Why am I in America as a successful scholar, a professor?*
3. *Why, out of fifteen girls in my fourth grade, was I lucky enough to move on to middle school, high school, and college?*

I successfully escaped the weeding system that began in the fourth grade, a level that determined the destiny of all the girls in my village. My mother and my aunts were not that lucky. Their education opportunities were cut short after the fourth grade, and they were forced to join their mothers on the farm and help in the house while they awaited marriage. My friends were weeded out one by one at each critical level—fourth grade, eighth grade, twelfth grade, and finally, after two years of college preparatory school. Despite a sizable number of us making it to college, only two of us had an opportunity to get a master's and doctorate degrees in the United States. I consider this destiny not luck. Every human being has a story to tell to provide answers to the core question, why me?

My story starts in my home country, part one of this book. In part two, I trace my journey from Tanzania to America where I have resided for over thirty years. My journey has not been without twists and turns, much of which I still do not understand. This is a place I never thought I would ever be. I am cognizant of the fact that a lot of my experiences are embedded with the love and respect of many of the people I met on the way. They helped me discover myself and discover a warm place away from my ancestral home.

The thought of writing this book lingered in my mind for many years. In 2018, when I finally retired, friends who had heard me talk about my childhood and my journey to America persuaded me to write a book about it. I resisted for a while, but when I realized I needed something to keep myself busy, I relented. The enthusiasm for starting the project led me to write a few pages, but it did not take long before I got involved in several assignments, including invitations to speak at conferences, which took away the writing momentum.

The COVID-19 pandemic and subsequent lockdown brought back the need to keep busy and not be depressed by the news of how devastating the coronavirus had become. It was hard to live in an environment where so much activity was curtailed, including such things as shopping, socializing with friends, going out to lunch or dinner, attending church services, or visiting with my family in Georgia. Writing became a way to keep me sane by avoiding the horrifying COVID-19 stories in the news. Waking up in the morning with a project in mind enhanced my plans for daily activities. In addition to writing, I decided to plant a vegetable garden, which served as a breaktime activity.

I walked around my garden checking the vegetables and chasing the squirrels. At one point I thought I had an army of squirrels attacking the corn and tomatoes. No deterrent seemed to work except frequent walks in the garden during my breaks. When the squirrels saw me coming, they ran out of my vegetable garden and climbed the nearest tree as fast as they could.

After two weeks of writing, I gained momentum and found it became a form of relaxation and an opportunity to relive my past. I had enough time to think about the story, putting it in sequence from my humble beginnings to life in America. I was combing through my life, trying to relive my childhood years, and walking different paths that led me to settle in the United States, a place I had read about in books but never thought I'd see in my lifetime. Whenever I finished a chapter, I saw the next one as a monumental task, and I was filled with mixed emotions. At times, it felt good to remind myself of my incredible journey, thinking about the different people I met along the way, and the valuable lessons and experiences that shaped my life.

My motivation for this book is to pay tribute to all who came before me, whose footsteps I have followed, and whose words of wisdom have inspired me and whose encouragement has kept me afloat and pushed me to do things for the common good. As I examined my journey from childhood to adulthood, I see, in my mind, the places I have visited or resided as a guiding light for my life journey. As a kid, I did not understand my behavior, but the exposure and experiences I have gone through since then have allowed me to psychoanalyze myself. The first realization came to me at the University of Dar es Salaam, when I was studying for my first degree, in a class taught by a female

American professor, Dr. Marjorie Mbilinyi. Before you start wondering about her name, which is not very American, she was married to a Tanzanian whose last name she took. Dr. Mbilinyi taught Educational Psychology, a class that introduced theory and practice through research. She sent us to community schools to follow one child for a week, documenting their social and learning behaviors. It was at this time I came to terms with my behavior from primary and middle school years, between the ages of six and fourteen. I was a firebrand during this period, acting out my frustrations and dislikes without thinking about the consequences. I spent a lot a time defying authority, speaking my mind, and defending those I thought were being treated unfairly. I did not understand the implications of a superimposed culture, an explanation I would like to offer for my defiance.

I thought about my experiences with the nuns even more when I watched the movie *Sister Act* with Whoopi Goldberg (1992). I liked that movie so much that I went to the theater three times to watch it and then bought the video. It was funny, but Whoopi's actions and reactions in the movie reminded me of growing up in the convent school. I saw the culture in a new light, as an adult and no longer a child. I could relate to Whoopi's reaction to the food, the rule of silence, and the motivation to break rules, as simple as picking tomatoes in the garden and hiding them in the pockets of her long convent dress to eat later.

Like Whoopi, I had trouble changing my cultural norms to those espoused by the English nuns who were educating us at a female-only boarding school. This clash of cultures got me into trouble frequently. The more I was punished, the more I resented the culture and the local staff as well as the African nuns at the school. I saw all the authority figures around me as the gatekeepers of the new culture, particularly the religious aspects of it, and I could not easily relate. In the absence of counseling at this boarding school, punishment became the necessary solution to everything the girls did. Most of what they interpreted as mischievous conduct was merely kids just being kids.

When I came to America, I started to understand the dilemma of how kids are labeled based on their behavior. I was resentful of labels and realized that even in America, kids resented being labeled. I became defiant until I found someone in this new culture who understood me and did not use labels to describe my behavior. My experiences over the years have enabled me to slowly learn how to overcome cultural and social conflicts. I realized in my education psychology class with Dr. Mbilinyi that it was not unusual for me to react the way I did in primary and middle school having grown up in a male-dominated family in a patrilineal society. In my early childhood, I learned I had to fight to survive social and cultural conflicts. I tried to defend those who I thought were vulnerable either from perceived injustices of the system or by bullies or others who tried to take advantage of others.

Besides documenting my life's journey, this book has been cathartic. It is also a gift of knowledge to my nieces and nephews in Tanzania and here in the United States. I want them to know me and my journey and to realize that I was an ordinary kid who wanted something better than what I saw around me. I was barefoot but had ambition. For those in Tanzania, I want them to know the details about my journey to America. I want them to embrace what changed our family history and is exemplified by my nephews and nieces who were born in the United States. They are the first generation of American Tanzanians in our family. They will have a story to tell their children and grandchildren. My story also will enable them to maintain their connections with their cousins and their children and grandchildren who might wonder why they are American Tanzanians.

— 1 —
THE DIRT COUNTRY ROAD FROM HOME

I was born in Tanzania in 1948 in a village called Kirua, to parents called John Kifaranja Semali and Martha John Sambonanga. The village is in a subdivision called Tella in the district of Moshi Rural North, about eighteen miles from the capital city Moshi, Kilimanjaro region. My parents had eleven children, nine boys and two girls. I am third in line after two boys, the eldest brother Wenceslaus followed by Ladislaus. My only sister Matilda was born right after me. Thereafter, it was boys: Frumence, Venance, Anthony, Joseph, Casian, Florentini, and Adelin, the last born. After the death of our mother Martha, in 1995, my father remarried and had three more children; Gundelinda, a girl, and two boys: Gervas and Augustino.

The village is 3,914 feet above sea level, about one-fifth the height of Mount Kilimanjaro, which is 19,341feet. Growing up, we could easily walk to the foot of this mountain, about six miles away. This is one of the spots where tourists can begin the climb of a lifetime leading to the peak. The area is surrounded by communities that owned houses and farms of coffee, bananas, and different cereals. For a person living in this area, the distance is negligible, one could walk longer distances to their farm and sometimes to the nearest town. Transportation by car or bus was limited. The closest walk was to school or church. As kids, we became accustomed to walking barefoot around the mountain since very few people owned shoes. Even when a child had shoes, they took them off in solidarity with those who could not afford a pair. We enjoyed looking at the peak as we skirted the mountain on a clear day. Sometimes we felt like Kilimanjaro was just a stone's throw from where we were standing, but in reality, it was miles away. The climate was mild, but it could be extremely cold with occasional frost. There is a huge forest surrounding the mountain, a sign of the abundant rainfall the region gets. Agriculture has always been the main occupation, mostly coffee, bananas, cereals like beans, maize, millet, and sunflower seeds in addition to a variety of potatoes, fruits, and vegetables.

This northern corridor is the home of Tanzania's natural resource treasures. The mountain is a major source of the country's foreign currency, attracting tourists from around the world who come to explore it and national parks including Arusha, Lake Manyara, Ngorongoro, and Serengeti. The Serengeti is the largest national park in Africa and is the only place in the world where tanzanite is found. During colonial times, this area was known as the Northern Region encompassing the Kilimanjaro, Tanga, and Arusha districts under one administrator, the regional commissioner. After independence, the new government broke the Northern Region into smaller regions.

The population of the Kilimanjaro Region is about 12,000 and with recent developments, most villagers have electricity, running water, and tarmac roads. This was not the case when I was growing up. We walked everywhere we went, and we depended on firewood for cooking and small kerosene lamps for light.

Because of its mild weather, coupled with the natural resources, missionaries flocked to the region. Initially, Germans, and later the British, settled here and built churches, schools, and medical centers that provided primary care health services while the main hospitals in town served as referrals. However, very few villagers depended on the modern health services, relying instead on homeopathic remedies. My grandmother had a vast knowledge of herbal medicine with a cure for every imaginable problem. She treated cuts, rashes, and wounds with leaves she picked from the gardens around her house. She crushed the herbs before treating the affected area and covered it with banana plant fibers to keep the flies away. We used to think that Grandma had healing magic. She could treat a small cut in the skin with herbs and reduce the healing process to hours. A bigger wound could heal in three days or less. Grandma knew when we were about to get sick. She believed that all illnesses started in the stomach. She scheduled body detoxing on Saturdays because we did not go to school that day. We were given dried seeds that I have no English or Swahili word for, but in my first language, Chaga, it is called *ngetsi*. Grandma dispensed a couple of tablespoons, watched us chew the seeds, and then wash it down with water. Another detox was green leaves that she pounded and boiled for half an hour. She would let it cool, added sour milk, and dispensed it. Grandma had a schedule for each child. It took about two hours to take effect. Once Grandma was convinced the stomach had expelled the unwanted matter, she would prepare porridge for the child. The hot porridge helped eliminate residues of the medicine from the stomach. In the evening, she would prepare green banana and beef soup to replenish the body's nutrients. Grandma oversaw our health when we were growing up until my mother and father took over the responsibility and introduced chemical medicines—mostly tablets for treating worms and castor oil or Epsom salts for detoxification.

My father had a coffee farm that the whole family worked on. Credit for the introduction of coffee growing in this area goes to the Germans who were the first to colonize the country. They used local labor to maintain the farms, but the chiefs were very smart and didn't allow the Germans to appropriate all the arable land for themselves. While they developed estate farms on allocated land, the chiefs divided the remaining acreage among their people, encouraging them to grow what the foreigners were growing. Some of the villagers worked on the estate farms, while maintaining their own little farms, harvesting coffee, bananas, fruits, and cereals.

The British inherited the administration of the country in November 1918, with full control of the region, its districts, and villages. Unlike the Germans who were more interested in industrialization, the British were keen on educating the locals to create a workforce for administrative and development jobs. After World War II, the British inherited the administration of these schools through missionaries. My father enrolled in the village school and later his children went to the same schools. The missionary's main objective was to convert the residents to Christianity.

My grandmother told me I was born a little before the cock crowed. By estimation, it must have been after 2 a.m. The timing is interesting and deserves a comment. If my grandmother is correct, my birthdate is either July 30 based on how time is conceptualized in the Tanzanian culture, or July 31 based on the way time is realized in the Western Hemisphere. In the West, the new day begins at midnight placing the "cock crow" timing on July 31. However, in the Tanzanian culture, there are twelve hours of daylight and twelve hours of nighttime. The day begins at six in the morning and ends at six p.m. Nighttime begins at seven p.m. and ends at six a.m. Thus, the official date on my baptismal certificate is July 30.

At my birth, my mother was happy to have me to add to the two boys because I was to become her helper around the house. In the culture, the birth of a child was an assurance to both the mother and her husband's family that she could have children. The birth of a boy was much more celebrated because the mother was assured a place in her husband's family, for she produced an heir to her husband's estate. The absence of an heir meant that when the man died, his estate would go to his brother(s), and the woman's security and wellbeing was at the mercy of the male members of the family.

My grandmother told me that my father was extraordinarily happy about my arrival. I did not have a chance to ask my father why, but I was my father's little girl and firstborn daughter. Grandma also told me that after my birth, my father left the house at dawn to go to town to buy supplies for his newborn girl. She could tell I was special to him. That day, my mother thought my father had left for school since he was a teacher in the village. Then, a few hours later, he returned with a bag of supplies, and my mother realized he had been shopping. It was the end of the month, so he had been paid his salary and had money to spend on his baby girl. He told my mother he wanted to get everything new. While having breakfast, my father remembered he forgot to buy a washbasin. After breakfast, he left the house again, leaving everyone to think he went to school, but he had gone back to town to get the washbasin. Finally, after accomplishing this task he went to school. He had sent word to the principal that he would be late. The principal kept his students busy until my father arrived around noon that day.

A week after I was born, I was baptized. My father had an interesting way of choosing his kids' names. He had a book of saints that he used to search for names that were not common in the village. For a long time, because there were no other children in the village who had my name; I thought I was the only one in the whole world with that name. But when I went to Europe, years later, I found more people with that name in England and Germany. I also learned the origin of the name was English and it was associated with a Christian saint who was born in England and later migrated to Germany.

My grandmother gave me my second name, Mkaficha. This is a compound name made of *mka*, "woman," and *ficha*, "special." The meaning is associated with a locally brewed banana beer. Banana beer is made by cooking ripe bananas for a long time until they turn a maroon color. They are left to ferment for a couple of days, and then the juice is squeezed out in stages. The first squeeze is done before any water is added and is reserved in a special container. Small amounts of water are added to brew different grades of the beer. The lowest grade is usually served last after all the better grades have been consumed. There are other ingredients added to the brewed beer such as ground red millet flour, which is cooked like porridge, left to cool, and added still warm to the liquid brew. The beer is left to stand overnight to be served the next day. This brew is typically served on special occasions like weddings, church holidays, and other community-based celebrations. The quality of the first brew is enhanced after the red millet is added and is considered special because only special people are served it, thus the name *ficha*. The beer is stored in a specially designed earthen pot made from local clay soil and is usually served a day or so later to close friends of the head of the household.

I did not know what my name meant until I was an adult and curious about who I was named after. Most second names are linked to a grandfather, grandmother, or event that was special to the parents or grandparent. As for my name, I only heard my grandmother use it occasionally. I suspected she did not really care for my baptismal name, which she found hard to pronounce because of the letter "b" which does not exist in my first language, Chaga. Any Swahili word with the letter "b" was substituted with "p" when introduced into the Chaga language.

Sharing responsibilities in my family was a must. The chores were divided based on gender with the girls spending most of their time with their mothers and grandmas. I spent a lot of my time with my paternal grandmother, Elizabeth—this was her baptismal name, but we knew her as Malya, the name she was given by her parents. She lived in the same compound with the rest of the family. My sister was sent to my maternal grandmother, Monica, where she grew up with my maternal aunts on my mother's side. As such, we did not grow up in the same household sharing the day-to-day activities, including babysitting my younger siblings. I saw my sister every Sunday at church or if my mother sent me over to my grandmother's house to fetch some yams or to take a verbal message to my aunts or grandma. I cherished the days my mother sent me over to my grandmother's house because I had a chance to spend some time with my sister.

I learned a lot from Grandma Malya who was at home more than my mother. She had small gardens, and whenever she went to work there, she took me with her. We also went together to look for firewood or cut grass for the two cows she kept for milk and manure. Houses were sectioned off to accommodate both people and animals. I learned a lot about her history and how she became my grandfather's fourth wife. Malya was Maasai, which was obvious from the big holes in her earlobes that held decorative beaded earrings. She made the earrings herself during the rainy season. My father did not allow my sister and I to have our ears pierced, so grandma made necklaces and bangles for us that we wore at home but not to school or church.

Growing up with grandma allowed me to understand her background. She grew up before the missionaries came to the region. Experiential and Indigenous knowledge were imparted through apprenticeship and nontraditional education provided by fathers to sons and mothers to daughters. My grandmother learned how to be a good wife and mother from her mother, and she passed this on to her daughter.

I do not know much about my paternal grandfather, Semali, because he died a few years before I was born. The little I know is that he was a determined man, just like my father, loving and fair to everyone in the village. He had four wives, and my grandmother was his fourth and youngest wife. He had an average of four children with each wife, but my grandmother had only two, my father and my aunt.

My grandmother was about twenty years old when she married my grandfather. She did not choose to marry him but was betrothed by her parents after an abduction. This happened one evening when she was cutting grass for the cows. Two men grabbed her and took her to my grandfather's house where she had to spend the night. She could not go back home the next morning because once a girl spent a night at a man's house, she was automatically declared defiled and had to marry the man. Her parents were comfortable with the stature of the family she was marrying into, and the wealth associated with landownership. They also knew that she would be given some independence with the choice of a homestead for her and her offspring.

Because my grandfather had three other wives with an average of four children each and two with my grandmother, my grandfather decided he did not need any more children. My grandmother had a different idea; she wanted her two children to have siblings, another boy, and another girl. She was infuriated that her co-wives had more than she had and might assume she was unable to conceive more children. Having a son insured her place in my grandfather's estate, and that was a relief. In a patrilineal society, sons were considered more valuable than girls; they were heirs of the father's estate. Girls were not considered valuable because their lives were claimed by the family they married into, even though their well-being was dependent on their ability to bear sons. My grandmother's security was assured, and the land allocated to her as the fourth wife could not be taken away from her because it belonged to her son who was her protector in this consortium of co-wives and their male descendants. To my grandfather's credit, not having more children, particularly sons, was strategic because of landownership rights. He would need more land to bequeath his sons and he wanted each son to have enough land that they could bequeath to their sons. Each needed enough land to farm and raise cattle.

Looking back, I am happy for my grandmother. She outlived her co-wives. My grandfather loved her and cared for her well. Until she died, when I was twenty years old, she stood tall, elegant, with little gray hair and glowing skin. She had not seen a doctor her entire life for any major illnesses, and her death came quickly. It is unclear what she died from.

Listening to my grandmother's stories and watching her and my mother go about their daily activities around the house and the farm did not inspire me to be like them. I sought an alternative. I was not sure what it was going to be, but I certainly did not like the idea of walking long distances to the farm, cutting grass in the woods for the cows, cleaning the animal stable, and walking on the cow poop without shoes. I liked my father's lifestyle as a teacher—get up in the morning and teach kids Monday through Friday.

My grandmother kept her small family remarkably close; they looked out for one another. My grandmother was very protective of my father, and, in turn, my father was very protective of her and his sister. This relationship shaped my father's sensitivity toward women, which was often displayed in private for fear of being considered a weakling in the community. My grandmother indicated that she was concerned that my father did not have male siblings to lean on, although there were other sons from my grandfather's other wives who shared paternity with him. Her fear was that the older boys would take advantage of him or show the same indifference as was shown to his father. This fear turned her into a woman with a mission—to empower him and raise him to be strong and independent. She did not think living on the farm and tending cattle was going to distinguish him from the other children, and although she did not believe in the white man's heavy hand on the village, she was willing to take advantage of the new product they brought to the village—education.

At that time, all the children went to school in the village through the fourth grade. When my father finished that grade, my grandmother decided to send him to a mission school despite my grandfather's objections. The objection had to do with school fees that my grandfather was required to pay so that my father could go to a mission school, selected for only a few of the students. When my grandfather refused to pay the fees, my grandmother stepped in and took matters into her own hands. She sold one of her cows to obtain the needed cash. My father left home for the missionary school and became a teacher. His experience became the foundation for his foresight about education, a benefit to all eleven of his children, especially the girls.

I was not aware of how resilient my grandmother was. I only saw her as Grandma, the woman who was always home with us when mother was out in the fields or sick or when she had a newborn. She shared parenting with Mother, a blessing for us when growing up. Having her house in the same compound was an advantage because she could watch over us and provide snacks or make dinner some days. She would boil maize with beans in a big pot that we could snack on all day. That was our equivalent of trail mix. We took some of it to school for our lunch break.

This family structure was the foundation and platform from which my siblings and I launched our successful lives. Both my mother's and my grandmother's kitchens were a continuous source of nourishment. No child was ever hungry, including our neighbor's kids. When my father bought meat from the butcher, both kitchens would get a portion. If my grandmother was cooking, my mother did not have to cook. Rarely did both cook two big meals at the same time. It was very well coordinated except that my grandmother did more of the snack foods than my mother. We enjoyed her snack foods and, most of all, her moonlight stories. The powerful moral lesson at the end of each story fascinated me and has stayed with me to date.

One of the stories was about a conceited girl who thought she was the prettiest and smartest. She was sent to gather firewood, accompanied by her playmates. On the way to the forest, one of the girls picked up a pebble and showed it to the others, saying, "Look how beautiful this pebble is. I am going to keep it in my pocket so I can look at it every day."

The conceited girl said to her, "Let me see."

The other girl pulled the pebble from her pocket and handed it to her. She looked at it, turned it around, examining its worth. She then threw it away and said, "It is the ugliest thing I have ever seen. It cannot be pretty like me."

The girl who found the pebble regretted showing the other the pebble and was not happy that her friend was disparaging the pebble and then threw it away. The girls went on to look for firewood for some time, and when they had collected enough, they tied it in a bundle and carried it back home on their heads. On their way home, they found a big rock blocking their path. They put their firewood on the ground and tried to roll the stone away but could not do it. The conceited girl told the others she was going to find a big stick to insert under the rock, and then roll it. While she was gone, the stone moved and a voice from the stone told them to take their loads of firewood and cross to the other side. Then the stone rolled back into place. When the conceited girl returned, she saw the other girls had crossed to the other side. They told her that the rock rolled to the side of the path and a voice from inside told them to cross. She called out to this mysterious voice to let her cross, but nothing happened. She started to cry. Suddenly an opening in the rock appeared. A voice told her to enter, and she did. The opening closed behind her. The other girls waited for a while, but she did not appear on the other side of the rock. It started to get dark, and the girls decided to go home without her. They told their parents what had happened, and the parents went to inform the girl's parents. The father and some men from the village went to the location to see for themselves. They found the stone and the girl's bundle of firewood. They called the girl's name but there was no answer. Because it was dark, they decided to go back home and try again in the morning. In the morning, the father, accompanied by the village leaders, went back to the location, and found everything they had seen the previous day. As they approached the rock, they heard the girl singing her favorite song. They tried to call her, but she kept singing. When she stopped, a voice from the rock said, "Do not worry, she is in good hands. She was very unkind to me, but I will not do likewise. For her to leave the rock, she must apologize, and she must pay ten healthy cows to our chief."

The father was happy to hear a conciliatory voice. They returned to the village to organize the ten cows needed. By that afternoon, they had all ten healthy cows and headed back to the location. The chief tapped on the rock to let the occupants know they had returned. The voice welcomed them, and they asked the girl to make her public apology. She did without hesitation. The voice inside the rock asked the villagers to line the cows in front of the rock and to move away one hundred feet. They did. An opening emerged and the cows went in, one by one. When the last cow entered, the opening closed. The villagers waited anxiously. Then the rock opened again, and the girl came out unharmed. She ran to her father who picked her up and put her on his shoulders. The opening closed and the rock disappeared. The villagers were astonished by the rock's supernatural powers. They returned to the village and told their wives to teach their daughters to be kind to all things human and non-human. The girl learned her lesson and was never unkind to anyone or anything since then.

The moral of the story was humility. There were many other tales, and each storytelling event was as exciting as the last. We looked forward to these moonlit nights and invited our neighborhood friends. My grandmother was everyone's grandmother, and she treated all the kids like her own biological grandchildren. Her stories made us enjoy radio programs when my father bought his first radio. All the kids would be glued to it listening intently to the actors. It was like watching TV soap operas, the anticipation of the next episode growing stronger each time. The only difference between the radio and TV programs and the oral stories was that each oral story was complete, and the moral was vivid with lasting effects on the listener.

Having Grandma at home also relieved me of the babysitting pressure. My responsibility as a babysitter came at age five. My hip bones were strong enough to balance a baby, with my hands providing a firm grip. Otherwise, I had the baby on my back, secured using a strong piece of cloth called *kanga*. Because of my age, I always had my mother or grandmother tie the *kanga* to ensure that there was no danger of it coming loose, causing the baby to fall off my back. All I could do with the baby on my hip or back was to stand around or sit down, watching the other kids play. With my grandmother at home, she would take the baby during feeding time or just to relieve me so I could also play. In this way, my grandmother was my best friend, surely a friend in need.

As kids, we used creativity in designing games and toys. We made our own toys using sticks, grass, and mud. We used anything we found lying around. The games were gender-typed. While the boys played soccer or had wrestling matches, the girls jumped rope or played the squares. The girls focused on family and organizational activities too, what we were exposed to daily. When the boys and girls played together, boys tended to assume leadership roles, ordering the girls around. When we played Mass, the girls prepared the altar and cut ripe bananas for communion while one of the boys was the celebrant (priest) or served as an altar server, assisting the celebrant. The girls formed the congregation and sang in the choir. But when it came to playing school, I was stubborn. I did not allow the boys to take charge. I thought boys were lousy teachers and they were less methodical and too mechanical. As a teacher, I taught my playmates math and English. I was very particular when we played school because I made sure that we had a schedule that followed the one we used in real life. The class time was no longer than the usual forty-five minutes, and we had fifteen minutes of play breaks. Little did I know that I would become a real teacher.

I started school early, at age six, and I skipped kindergarten because I had learned a lot from my older brothers and from playing school at home. Since my father was a teacher, there was a learning routine in the house. As kids, we had homework time after dinner. This was also the time my father prepared for his classes for the next day. He gave the boys math problems to solve as well as writing exercises. I watched and practiced with them. One day I took my father's pen when he was not at home and practiced my writing in the book he used for his catechism classes. While I was pleased with my creativity, my father was not amused. He decided then it was time for me to go to school to put this energy to work.

I liked school, but I was a handful for my teachers. I liked Friday mornings the best because all the kids in first through third grades participated in a music class. I liked to sing, and if the song appealed to me, I started dancing. The music teacher did not discourage me, and the more the students cheered me on, the harder I danced. I do not think the music teacher told my father, who taught at the same school, because he would have told me to stop acting up.

In my third-grade class, I found myself under my father's watchful eye. He taught math, and I was good at it, finishing the exercises as soon as he finished writing them on the board. To keep me busy and less mischievous, he assigned me a struggling student to help. His class was usually in the afternoon after lunch. My father pretended to nap, with his head on the desk, but he was tricking us to see who was misbehaving in his class. One day, I finished helping the student and started making funny faces that made the kids giggle. My father called me to his desk and asked me to kneel right next to him facing the class. When we got home, I received additional punishment for disrupting his class. The incident made me consider moving from my father's class to another one. It was too much to be under his watchful eye both at home and at school. After a few days of thinking, I decided to go to another class taught by a nun from the Order of the Sisters of Mount Kilimanjaro. I told her my father was a lion and I did not want to be in his class anymore. First, she laughed and then asked me, "How about your mother?" I answered emphatically, "She is a sheep." She allowed me to stay and informed my father that I had transferred to her class. I though my father would ask me about it when we got home. For two days, I avoided coming in close contact with my father to reduce the chances of him asking me about my actions. It was just my imagination. He did not ask me, but he told my mother about it.

In third and fourth grade, I was the class monitor, leaving no room for me to be mischievous. I enjoyed the leadership role and was very protective of the introverts. One day, during recess, I saw a boy bullying another boy from my third-grade class. He was significantly smaller in height and size. I approached them and listened to what was going on. Once I determined that the bigger boy was being malicious, I took matters into my hands and told him not to bother the little boy again. The bully was bigger than me. I was medium weight and unusually short. I should have been afraid of him, but as a class monitor, I thought I could expand my jurisdiction to the playground. I looked fearless, which surprised the bully. With my arms akimbo, I approached him as close as I could to show him that I was not afraid of him. I told him that if I saw him bothering the little boy again—or anyone else—I would hang him on a tree that stood imposingly in the school compound, and I would let the hawks pick at his body, piece by piece until he was dead. Of course, I did not have those powers, but the poor bully believed me and took off. He told his class teacher that I had threatened him, but he did not tell his teacher what had transpired before the threat. The teacher told my father about it. My father promised the teacher that he would investigate it when he got home. To please the bully, the teacher informed him what my father had promised to do. Thinking he won, he saw me after school and shouted, "Your father is going to deal with you when you get home."

I was not afraid, and when my father asked me about it, I told him that it was true and that I had threatened the boy. My father had not been told the reason. I told him the story and he looked at me, smiled, and then said, "It is important to protect those who are unable to protect themselves. Try to stay out of trouble."

I nodded and realized I was not being reprimanded but was being encouraged to do good. The next day, I looked for the bully during recess. I went up to him and said that my father knew what he had done, and I meant what I said. I would not let him bully anyone on the playground. He took me seriously and never crossed my path again.

In fourth grade, we were introduced to single-sex classes. The boys were taught by male teachers and the girls by females. There was an all-boys middle school across the street where the boys could go after fourth grade. The girls had to go to a boarding school away from home. There were four boarding schools for girls in the entire region, two for girls from Catholic primary schools and two for Lutheran schools. Such restrictions made it hard for the girls to go beyond fourth grade while all the boys were almost guaranteed a place in middle school. If a boy did not get in, they were given a chance to repeat fourth grade to try and be accepted the following year.

As fourth graders, we did not realize what the stakes were. At the end of that year, we took the exams as required. Everyone in the boys' class passed and moved on to middle school. None of the girls in my class made grades that were competitive enough to advance to the middle school run by the Assumption Sisters from England or to the second one much farther from the village and run by the Catholic Church dioceses. It was heartbreaking to all of us and more so to our teacher who was reprimanded by the school principal. She was moved to third grade and that was the end of an all-girls fourth-grade class. Then, something interesting happened. Instead of all of us being sent home because we failed, it was decided that we would repeat the fourth grade. This was because two of us in the class were daughters of teachers at the school and one of these teachers was the principal. My father and the principal were great friends and played soccer every Friday. They also visited each other's homes. I was friends with his daughter and now both of us were in this predicament. Our parents were not ready to send us home to the lives of our mothers. They had to do something that had to include the entire class because it would not look right to just do something for two out of fifteen girls. The principal decided to have all the girls moved to the boys' fourth-grade class making it a co-ed class. The teacher for this class was also the principal. He was determined to rescue the girls, something that earned him a lot of praise from the female teachers at the school as well as the parents in the community.

Little did we know we were going to have a year like no other. He was determined to whip us into shape, to make sure we worked hard enough to get the required grades to go to either of the two middle schools available to girls from the village. He was very condescending, telling us we moved too slowly and worked too slowly on the math problems. One of the things he did each day was write ten problems on the board. He required the students to write answers in their math exercise books at the same speed he was writing on the board. This meant that when he put his chalk down, we too had to put our pens down and declare "finished." Many of the boys could do that easily but not all the girls. Some girls waited until he was done writing and then asked him if they could start writing the answers in their books. They did not understand his instructions were definitive and not circumspect. When he became furious with the girls, he would call them names. One day after class, some of us met to devise a strategy to get out of this problem. We resolved that we would change the situation by facing the challenge. We would race the boys to the finish line and beat them. The next time the teacher wrote the questions on the board, he thought he would be chastising us as usual, but several of us put our pens down at the same time as he did his piece of chalk. He turned to the class and mockingly said, "I bet the girls are still stuck and thinking about pounding millet."

I was furious and tired of him putting the girls down. I looked around and saw several girls had finished writing. I saw some boys still writing as well as some girls. I thought we were even on this one. I decided to respond to his claims and said, "Of course not. Don't you see some of us are already done?"

The principal did not like me standing up to him. He walked toward me and before I knew it, he slapped me, leaving his five finger marks on my cheek. He then called out to all the students to stop writing and asked the class monitor to collect the exercise books and take them to the staff's common room. I held my tears in defiance, and I could see all the girls had their heads lowered, looking at their desks and not the teacher. It was break time, so he dismissed the class. I went straight to my father's classroom and showed him the marks on my face. I do not know what I wanted him to say or do, but I wanted him to know what had happened to me. He listened to my story and then told me to go join my friends on the playground. Of course, he did not do anything because he did not want to anger the principal. Picking a fight with him could have resulted in my father being transferred to another school, far away. At that time, the school was only a ten-minute walk from home. Later that day, when I got home, I told my mother about it. She examined the bruise and assured me it was going to be fine. She took a wet cloth and applied pressure on the affected area. By the next day, the swelling was gone. She also told me to avoid vexing him and focus on my work regardless of his comments. It was hard to take, but I promised her I would do that.

The rest of the year went well. We had our exams as usual and had to wait for a month to know our fate. The exams were set by the regional Department of Education, and when they were released to the schools, a supervising teacher was picked for another school and district in the region. After the exams, the papers were taken to the district office and graded by selected examiners. After that, the district decided which of the students did well enough to go to middle school. The boys had an advantage because there were ample middle school in the different villages where they could be placed after the fourth grade. There were only three middle schools for girls in the whole region, two were run by Catholic missionaries while the third was run by missionaries of the Lutheran church. At that time, religious affiliation determined where a child could go to school.

Every day before school, I went to church. At our house, unless one was sick, going to daily Mass during the week was a must unless there was a compelling case to skip it. Mama went occasionally, but most of the time she stayed home to get breakfast going. Because church and school were close to where we lived, we could run back home, eat breakfast, pick up our lunch packs, and then run back to school. I liked going to school to avoid house chores. I wanted a different experience than my mother, my grandmother, and aunts on both sides of the family. I knew from early on that I did not want to live their lives, which had no job description and work hours, but were full of endless sacrifices for others. My mother did so many different things in a limited number of daylight hours. Her day started at five in the morning when the cock crowed and ended late at night when we were all fast asleep. I used to tell my friends that when the cock crowed at our house it was saying "Good morning Martha", urging my mother to get up and begin her day of endless chores. I did not think there was anyone else in the village who woke up at the first cock crow. One day, the cock crowed at four in the morning and my mother got out of bed. On her way to the kitchen, she looked at the clock on the wall in the living room. She then realized that the cock must have gone nuts. She came back to bed muttering to herself, "You stupid rooster, you woke me up too early." She tried to go back to sleep but couldn't. She decided to go to the kitchen anyway to get a head start on her daily chores.

I thought my mother was a superwoman; it was hard for me to keep track of all that she did. As a result, I was a poor informant for my father when he came home from school and asked me, "What did your mother do today?" My answer was always, "A lot of stuff." I figured that if I tried to enumerate, my father would lose his patience and walk away in the middle of my partial list. I was also worried that I would sell my mother short because it was easy to overlook some of the activities. In addition, I was always not with my mother. If I had a chance, I sneaked out to my neighbor's house to play for a few minutes before the baby woke up and start screaming.

Yes, babysitting was one of my responsibilities. I was like a second mom from age three. I had to learn early how to balance the baby on my hip and to bear the baby's weight. Otherwise, I would carry the baby on my back, tightly secured by a *kanga*, a piece of cloth usually worn by women, and which looked like a big scarf. I would ask my mother to help make sure it was tight enough over one shoulder and under the other arm, a crossbody tie. If my mother or my grandmother was not available, I would ask my playmates to securely hold the baby on my back while I tied the cloth as tightly as I could. Practice made it perfect over time. My freedom was school. I could play all I wanted without being responsible for an infant or toddler. As soon as I got back home from school, my mother would hand me the baby to allow her to focus on making dinner and other activities around the house.

I had other chores, like going to the river or the natural spring to get water, gathering firewood, helping mother or grandmother prepare ingredients for the evening dinner, and most often, cutting grass for the cows. Each household had at least two cows and maybe a calf or two. Apart from being a cultural symbol, the cows were a source of milk for the family and manure for the garden. Livestock was an asset—a cow, goat, or sheep could be sold to raise money for family needs. Taking care of the cattle was a shared responsibility between males and females of the household, but the women assumed a larger role in cutting the grass, milking, and cleaning the stable. These cattle did not go out to graze. Instead, they were fed in place for their entire lifetime. Because my mother and my grandmother learned this when they were young, they made sure my sister and I knew that this was a responsibility for us as females. By age four, we were supposed to know how to balance a small can of water on our heads, babysit, and cut and tie a buddle of grass for the cows in the house. I thought school was the only escape.

This responsibility loomed large in my mind as I awaited the results of my fourth-grade examination. I wondered if I would pass, after what happened the year before, to forge a pathway to middle school. The exam results were a ticket for me to the new world that I was about to experience, though with apprehension because of the unknown. If I passed and got a place at one of the middle schools, I would be leaving home for six months at a time.

By now I had realized that my mother's life was different from my father's. It did not sit well with me when my mother told me that they were in the same school when my mother was in first grade. While my father went on to middle school, my mother's father pulled her out of school before fourth grade and was made to stay home and help her mother around the house and farm. My mother was the eldest girl in a family of three boys and three girls. I thought about the teachers at my school and the possibility that my mother could have been one of them if she had had a chance. She was very smart and helped us with our homework when she had time.

During my mother's youth, many girls were pulled out of school before fourth grade, but their brothers went on to become successful government employees. One of my aunts got married and then died in childbirth. The baby did not survive either. The other aunt married later and then was abused by her husband and his family and died alone at home. Before she got married, I helped her find a job at my high school. She worked very hard and in appreciation, the nuns promoted her to overseeing school maintenance. She worked until she was able to retire, a blessing because she had an income to sustain her in her later years. Her husband died a few years later, and my aunt remained in the house alone. A week before she died, she had collected her monthly retirement pay. Someone must have known because it was stolen from her house. My aunt came down with a stomach virus and because she was not able to go to the hospital, she died from hunger and dehydration. She was found three days after her death.

What my mother and my aunts went through is not unique. There were many examples in the village of girls like them. While my father was encouraged to go to school and was supported by missionaries to pursue a career in teaching, my aunt was not afforded that privilege. My grandmother was keen on her son getting an education even though she never went to school. When my grandfather refused to find the money for my father's school fees, my grandmother sold one of her cows to help my father meet his educational needs. My aunt remained in the house to help my grandmother around the house and on the farm. Hearing their stories made me aspire to be more like my father than them.

As a teacher, my father left the house around 7:15 after the morning news on his radio. Yes, it was his radio because he was the only one who turned it on and off. There was no radio if he was not at home. Years later, this inspired one of my little brothers, Anthony, to make his own radio using electric wires he found at a construction site and used batteries that he recharged by leaving them in the sun for a long time. I thought he was a genius and could not understand how he pulled it off.

To get to school in the morning on time, we made sure we left at the same time as Dad. When he was not the teacher responsible for supervising the weekly activities, we left before he did. We did not want to be late when he was that week's supervisor. There was no feasible explanation if you arrived at school a second after the bell for assembly had rung. You could not sneak into the already neatly formed assembly lines of students, separated by class and gender, and waiting for the school principal to emerge to make daily announcements. All the kids knew that my father's eyes were like those of an eagle. He could see far and from both sides without turning his head. Because he was on schedule, everyone else was expected to be on schedule.

At home, some things were relaxed. My mother had no schedule for her work or our chores. Dad had some—like when to wake up, clean the house before school, get to school on time, come home on time, get feed for the cows and goats, fetch water, have his water boiled and ready for his bath, especially on Fridays when he played soccer. If he needed it on other days, his request took precedence while other things had to wait. This was a way to instill discipline in us. Dinner time also was scheduled, unless he was not home. We ate around seven o'clock, and then the girls cleared and washed the dishes, and artfully arranged them in a tiny cabinet in the small dining room/family room. By eight o'clock we were done, and it was time for schoolwork. My father took out his books and started making lesson plans and notes for the following day. Because we did not take schoolwork home—all homework was completed at school—Dad would give us math problems to solve. This would go on for at least an hour before we were dismissed for bed. We had kerosene lamps that were extinguished soon to conserve oil. The boys stuck to this schedule, while the girls (only my sister and I) could vary it because Mother's routines sometimes interfered. Babysitting or helping in the kitchen could not be scheduled; it just happened.

Schedules were altered during the coffee-picking season. Coffee picked during the day had to be pulped the same day and put into special containers to ferment for three days. Depending on how much there was to pulp, the entire evening could be spent on this activity. Dinner would be served afterward and then we would all go straight to bed because everyone was so exhausted. The coffee-picking season lasted from June to October. It was extremely hard on us because the process was tedious. It often coincided with the harvest season when we had to walk four miles to the farm to harvest beans, millet, and corn. All this had to be hauled back to the house and prepared for storage. Sometimes we had help from people my father hired, but our family did much of the work. We had to carry the harvest on our heads from the farm to the house. This was mostly done on weekends. Coffee picking, on the other hand, happened whenever the berries looked ripe for picking. At the peak of the season, we would come home from school and head straight to the coffee farm where my mother had been all day. Sometimes her neighbors would visit and help her for a short while. Other times, her sister would help, but by the time we got back from school, they would be gone. She was always happy to see us since we broke the loneliness she endured on the farm as she moved around the trees laden with ripe coffee berries. Dad joined us as soon as he got home and, all together, we would move quickly in between the trees trying to finish picking the area before dark. Then, we would start on the pulping process, which took three or four hours. Saturdays and holidays during the coffee season were spent picking and/or processing the berries.

We got used to having our regular schedule disrupted during the harvest and coffee seasons. The grains provided most of our staples, and coffee was our main source of income. Proceeds from the sale of the coffee allowed us to have clothes, a new roof, or an additional room built in the compound. Later, it paid for our tuition as we progressed through middle and high school. The only time parents did not have to pay tuition was when their kids were in primary school, K-4. In my father's case, the family was growing fast, and he had more mouths to feed, clothes to buy, savings for tuition, and adequate shelter.

We needed a good house because of the weather. Being so close to the mountain, the area was extremely cold, and sometimes frosty and rainy. The rainy season was like snow days in Europe and America. The only difference is that in Europe and America you could attempt to drive someplace. It was impossible to walk or drive in the village during the rainy season because it was so slippery. It was easy to fall and no fun because the red clay stained clothes badly. Doing laundry was a test since it was done by hand and sun-dried. With no sun, it took days for clothes to dry. We were fortunate to have two uniforms that allowed a week's wash and wear. In the house, we had a storage area with a couple of clotheslines. In the winter it was full of wet clothing waiting to dry. Because it was connected to the bedrooms, the body heat that circulated through the house helped dry the clothes. We had to take care of the animals too and prepare our own food. We cut a lot of grass for the animals and kept it in a makeshift stall to protect it from the rain. We also had to stock up on firewood for cooking.

I liked the long rainy season because that was the only time Mama and Grandma were home ninety percent of the time. One of the things I hated most was getting home from school to find that neither was home. I could tell they were not there if there was no smoke coming from the kitchen chimney. I knew there would be instructions from my mother to start dinner or clean the animal stable or to meet her at the farm to carry home things like grass for the animals, firewood, or harvest that we were going to use for dinner that night. As soon as I got close to home, my little brothers would start naming the chores on my mother's list. If there was smoke from my grandmother's chimney, I was assured she was making dinner and my burden would be lighter.

 I learned a lot of valuable things from my grandmother. Both she and my mother did beadwork, making bangles and necklaces. My grandmother made earrings too, but she was the only one who wore them because she had pierced ears. My father allowed us to wear necklaces for church but disapproved of earrings and bracelets. I also helped my mother and my grandmother dehydrate green bananas. Whenever we had a banana bumper crop, we spent days peeling them, splitting them in half, and drying them in the sun. When completely dried, we stashed them in big sisal bags and put them in dry storage. These became handy when banana production was low, and we were waiting for the corn harvest. It was interesting to watch my mother and grandmother make our favorite banana dishes using the dried bananas. First, they boiled water and poured it into a container with the dried banana, letting it stand until the water cooled and the dried bananas came back to life, looking like they had just been peeled. They could use them in dishes like banana and beef stew or mashed bananas that could be eaten with meat stew, beans, or other vegetables. One of the interesting ways they used the dried bananas was to grind them with maize, making flour. This flour was used for a dish called *ugali*, which is usually made from maize flour only. Mixing a small amount of maize flour with the banana flour stretched the maize supply. The dish had a unique taste, better and softer than when just made with maize, especially when eaten with homemade yogurt. It was a lot of fun learning how to milk the cows and make yogurt and butter, but those skills have been mostly abandoned. Society now has grown dependent on imported processed foods, a phenomenon blamed for the rise of catastrophic illnesses like high blood pressure and diabetes.

— 2 —

The Chosen Few

Even though I enjoyed learning about the traditional and cultural aspects of life from my mother and my grandmother at home, I liked going to school more. It was a good escape, even if just for a few hours. I could be myself, be a kid, even at the risk of doing something at school that resulted in a reprimand or punishment. It was social independence, the kind of freedom kids longed for during COVID-19. They wanted to go back to school to be with their friends. They wanted to claim their time and space and be under the watchful eyes of parents and siblings.

I also had another reason for wanting to go to school. My father had an advantage—he was able to control everything he did while he delegated some things to others. I liked his self-determination and independence. I could not reconcile my mother's place in the grand scheme of things. I told myself that when I grew up, I would like to be like my father.

THE EXAM RESULTS ARE IN
It was Friday, the last day of the school week. Usually, Fridays were reserved for cleaning around school, and all the students had assigned areas. The class monitors were busy supervising, and the students were equipped with the tools

they needed to accomplish their chores. These included brooms made of small tree branches or special shrubs, a small watering can, and a hoe or sickle for cutting grass. Each student was required to have one of these implements, and if one showed up at school without them, they were sent back home to retrieve them. The school grounds needed to be swept, the classrooms and teachers' shared office space were to be cleaned with water and a wet cloth, and all the tables, desks, and windows needed to be dusted. Some students were sent to clean the bathrooms, others went to collect trash around the playground, and a small group was sent to help the parish gardener weed the vegetable garden. It was almost ten o'clock when the first bell sounded. This was a signal to stop working, wash up, and gather for the principal's address at the school assembly.

Being a Friday and near the end of the semester, school was going to be a half-day. We were also closing the school year, so the end-of-term exams had been completed. The teachers were still busy grading while the students waited anxiously to find out if they could advance to the next level. The half-day was added time for the teachers to finish their grading. It was also a sports day, and after lunch, we went to the playground to watch a soccer match. This Friday, it was the middle school across the street versus our school's fourth-grade team. Everyone was ready for this match. It was a good way to keep worries about exam results at bay.

The fourth graders' anxiety was different from that of the other kids because their exam results marked the end of a line and a time to exit primary school. We were all wrapped in a cloud of uncertainty. This was a make-or-break deal, more so for the girls than the boys. If a girl passed the exam, she had an opportunity to leave the village and go to a boarding school. Being in a Catholic school, I was hoping to pass and go to the new school run by the Assumption Sisters from England. The school had only four grades, fifth through eighth. Although it was a private school, it was regulated by the regional education commissioner and supervised by the district education officer. Per the rules established by these two offices, each class could accommodate only thirty-five girls from the various primary schools in the region that was under the auspices of the Catholic Church. In this case, more than one-thousand girls from across the region were vying to get in. The lucky ones would continue to middle school while the rest would begin the long road to motherhood and life in the village.

There had been rumors that the exam results were in and that the school principal would announce them before the end of the school term, which was only in a few days. It was the last week of November and the last week of school. The school would go on recess until January 7, when the new school year began. Although it was inevitable, we were not ready to exit school life, which was more disciplined and scheduled. We were only eleven years old. There were four female teachers at my school. I particularly liked one of them, Catherine. My father had once told me to follow in her footsteps. At that time, I thought my father was referring to her upbringing, politeness, kindness, and good reputation in the village. I had not seriously thought of her as a career role model.

The teacher blew the whistle, a sign that every student should be in line and quiet as the principal arrived at the assembly grounds. Shortly after, the principal emerged, and all the students said good morning in unison. A brief silence followed. Then the principal commented on how hard we had been working to prepare the school grounds and the classrooms for the end of the year recess. Then he opened an envelope and said, "I have been informed by the district education office that our school did very well in the fourth-grade examination. All the boys have been accepted to join fifth grade in our middle school in this village."

He proceeded to call out the names of the boys, pausing after each name to allow the audience to clap. The principal then said, "Okay, now the results for the girls."

All the girls froze in place. The principal continued, "I am happy to announce that thirteen of the fifteen girls will proceed to middle school. Twelve will go to Mandaka Middle School and one to Kibosho Middle School."

He called out the name of the student who was going to Kibosho Middle School. This was the school run by the Catholic diocese and was located on the western slopes of Mount Kilimanjaro. We all looked at the girl and smiled, made the sign of the cross, and raised our clasped hands to the heavens in thanksgiving. The tension grew among the remaining fourteen girls. Then the principal announced, "Now I am going to read the names of the twelve girls going to Mandaka Middle School, run by the Assumption Sisters from Kensington, England."

Each of the girls grabbed the hand of the girl standing next to her. I am sure we were all saying a quiet prayer, "Lord, do not let it be me." The whole compound was so quiet you could hear a pin drop on the dusty ground. My father's eyes were looking straight up, not at me. I did not know if he knew anything. I began to worry about going home that evening if my name was not one of the twelve. I started planning my escape to my maternal grandmother's house where I would get the sympathy of my aunts. Then the principal started calling out the names. I was getting worried when he reached number seven and my name had not come up. Then he called number eight and gave my name. The girl standing next to me squeezed my hand, and I placed my head on her shoulder. I was relieved; I was in. The principal went on to name the remaining girls. There was a big sigh of relief from each as the principal read their name, but nobody dared make a loud cheer. We were in solidarity with each other because we knew how hurt anyone left out would be. We could see that we had been robbed of the joy enjoyed by the boys. For them, there were seventy desks to be filled compared to only thirty-five for the girls. We did not have the statistics then. It was grossly unfair, but the system was set up unfairly from the start by the missionaries who did not think it was important to educate girls. The system was corrected after independence in 1962 when the first president of Tanzania, Julius Kambarage Nyerere, decided to establish day middle schools for girls in every village but left the private schools intact. This meant that any girl who did not make the favored list for boarding school could continue her education in the village.

When the principal finished reading out the names, the rest of the students and the teachers clapped, but there were no loud cheers like there had been for the boys. The two girls whose names were left out started crying uncontrollably. All the girls surrounded them, trying to console them, and at the same time crying too. It was unclear whether these were tears of joy or sorrow. I would like to think it was both. I was relieved that I would not have to run away from home to my maternal grandmother's house, but I was sad to see how hurt the two girls were.

The principal called the school back to attention and concluded by noting that he was pleased with the results and wished all those who had passed good preparation for their new school. He turned to the new fourth graders and reminded them that working hard brings good results. He then concluded the assembly. For the thirteen girls, this day brought both jubilation and sadness. We could not see ourselves jumping up and down with joy like the boys because two of our girls were hurting. We wanted to show empathy and share our solidarity with them, but there was nothing we could do or say that would ease their pain. We wanted to go back to our classroom, but before we could move, the female teachers at our school came over to us. They shook hands with all the girls, saying congratulations and good luck. They held each of the two girls who were sobbing uncontrollably, consoling them, and telling them it was going to be fine. All the girls were crying, a mixture of joy and sadness. This must have moved the teachers who were like mothers and big sisters to all of us. I was not sure what they meant by "it is going to be fine," and I wondered if they had a plan up their sleeves. What they said next assured me of it. The teachers tried to explain to the two girls that they did not fail, and they knew it was unfair and were sorry that the boys had more space allocations than the girls. We all knew that no amount of consolation could ease their pain. but the teachers promised they would help them submit a request to repeat the class and to retake the exam at the end of the new year. I walked back to class with the two girls and said, "You heard them, do not worry. It is going to be fine. They promised they are going to help you."

The teachers kept their word, and at the beginning of the new year, the two girls were called back to school to repeat the fourth grade. At the end of that year, they were selected to go to Kibosho Middle School.

Our teacher returned after we were all settled in the classroom. He had taken off his hat as school principal and now was talking to us as our teacher. He spoke about our new schools, admonishing us at the same time.

"You are going to be a few among several students from other villages. I do not think they are as smart as you are, but I want to warn you, do not bring me shame; do not bring my school shame. Iwa Primary School is the best in the region. You must maintain that reputation for us so we can send more students where you are going."

We sat quietly nodding our heads as he spoke. He also announced that teacher Catherine had been asked to join the staff at Mandaka Middle School. He noted that we should be happy to move on with her and have one of our teachers to go to if we had any problems. I was thrilled. My favorite teacher was going with me to the new school. When I was in second grade, Catherine knitted two sweaters, one for me and one for my sister. Mine was black with yellow stripes across the hem, the chest, and around the sleeves. My sister's sweater was green with the same yellow stripes. It had been my favorite sweater, and I wore it every day over my uniform during the cold season. It was big enough for me to wear through third grade.

We left school that day with mixed feelings but mostly relieved. Knowing that I was advancing to middle school assured me of a different path, one that my mother, grandmother, and aunts were denied. At home, I told my mother about passing the big exam and that I was going to a boarding school. I do not remember if my mother showed any emotions about it. I am sure she was trying to let it sink in, coming to terms with my leaving home at age eleven. At dinner, my father announced to everyone, "This one will be leaving in January to go to boarding school."

I know he was happy and relieved, but he did not show any emotion about it either. It seemed like just another day, a new event and off we go.

Saturday had arrived, a time when everyone was at home if there was no farmwork. It seemed like a typical Saturday to me after the great news, yet different. I could not stop thinking about what had happened the day before. The changes accompanying this transition were beginning to weigh on me. I had a lot of questions and didn't know who to ask because I did not know anyone else who had gone through these experiences. I also was going to be the first child to leave home. My two older brothers, Wencesalus and Ladislaus were not scheduled to leave until high school, in four more years

This Saturday, I helped with spring cleaning the different rooms of the house, doing laundry, and helping my mother with whatever she needed. My sister was still living with my maternal grandmother, so I was my mother's only helper. The boys had their own chores, and unless my father was out of the house, they could not be seen hanging out in the kitchen area. My mother could send them to fetch water from the river or natural spring or to look for firewood, but my father prescribed most of their chores, which included taking care of the goats.

Laundry included washing some of my father's clothes, although my mother did most of it during the week, my bedding and that of my younger brothers, window curtains, and tablecloths. Sometimes my mother would ask me to wash some of her clothes, but that was rare. The hardest was washing my little brother's school uniforms—white shirts, and khaki pants. It took me a long time to get out the red clay stains. My mother taught me a trick—soak them in soap suds and wash them last. This trick worked, but often I had to soak them for a long time and use a corn cob to scrub away the stains.

This time of year was slow because most of the work at the farm was completed until January when cultivation started again. Coffee picking season also was over. But some of the coffee would still be drying up in the attic. Occasionally, my father would bring down bags of dried coffee and spread the beans on a piece of wire mesh held together by four pieces of wood to form a square. This tool is known as *cheke cheke* because it looks like a square wire-meshed sieve. This tool is versatile because it is also used during the early stages of processing the coffee after fermentation, when the coffee has been thoroughly washed to remove the slimy texture on the beans and then dried in the sun before being transferred to the attic. When the coffee was brought down from the attic, we knew that the processing had come to the last stage before my father would take it to the cooperative to sell. The last step of the process included sorting the coffee into different grades. We had to look for perfect beans, which were fully husked with no cracks. These were designated grade one. Those with small cracks were grade two, and any with multiple cracks or that did not have a healthy bean inside were designated grade three. We would put the beans in bags specifically designed for coffee. Each grade had a different price. My father tried to have a bigger supply of grades one and two. The amount designated for grade three was too small, so my father reserved it for himself to roast, grind, and make a cup of coffee whenever he wanted one. He was the only one in the house who drank coffee; children were not allowed to drink it. We never asked why; we just drank tea. I never saw my mother or grandmother drink coffee either. I thought it was something only fathers did, and I was okay with it.

Although we still had coffee in the attic, this Saturday was not designated for the sorting process. So, it was a slow Saturday with just the usual chores that got us prepared for Sunday. As I went through my routines, I kept thinking about what was going to happen when I was gone for six months. I wondered if all that I did was going to fall onto my mother's lap or whether it would be time to bring my sister back home. I felt sorry for my mother, but at the same time, I hoped that my father would now change his outlook and let the boys assume some of these responsibilities.

I finished my main chores by lunchtime. Lunch on some Saturdays was special. My father liked liver (cow). He bought at least four pounds, which my mother prepared using a secret recipe that I learned. I have yet to find tastier liver. She also roasted green bananas as a side dish. Because I was slightly anemic, my mother used to give me more liver than my siblings received. My father gave me some of his share too. My father was in a good mood this Saturday and bought liver for lunch. As usual, my mother did a fantastic job. I was favored again, not solely because I was anemic but, I think, to celebrate my accomplishments. As a ten-year-old, I was fine with being treated special for the day.

My mother had a few more chores to complete before she started with dinner. I accompanied her to her garden, which was about a quarter-mile away from the house, to get yams she was going to use for dinner. She sent me home to start peeling them. She also instructed me to peel bananas she had left in the kitchen. Meanwhile, my mother remained in the garden area to cut grass for the cows and collect some firewood. She needed enough of both because the following day was Sunday, the day spent at church and doing homebound activities. I hurried home, peeled the yams and the bananas, and put them in water to soak. I went to the river to fetch some water and then embarked on ironing some of the clothes I had washed, focusing on what we would need for church the next day. Whatever was left undone could wait until Sunday evening, and this would be mostly school uniforms.

I was looking forward to Sunday because I knew I would see the eleven girls going with me to Mandaka Middle School. I was already imagining what life would be like. I wondered what the other girls were thinking. Then, I thought of the two girls whose aspirations might come to a sudden halt at age ten. Why them and not us? This was a troubling question I could not honestly answer. I tried to block it from my mind by looking for something else to distract me.

Seeing the other girls at church that Sunday morning was a thrill. Walking back home after church, we were loud and occasionally jumping up and down while screaming at the top of our lungs, typical adolescent girls. We were imagining what was ahead of us at our new school. We could not avoid feeling privileged and certainly not entitled to this success. One of the girls mentioned the two girls who did not make the list.

"Do you remember that short boy who usually sat near the window?" she asked.

"Oh, the one who never got the sums right. Teacher B. was always on his case," another girl noted.

"Yeah, that's the one. What I want to know is how he could pass to go to middle school when Mary and Angelina who performed better in class could not. I don't get it," she said.

A brief silence ensued. Then another girl commented, "Girls, there is nothing we can do about it. Remember what the teachers told us. Mary and Angelina did not fail, it was just that the places reserved for girls are fewer compared to the offers given to boys. The commissioner would assign any dumb boy to the middle school to fill the slots while some girls, who might be smarter, were left out."

Another girl picked up the conversation. "Well, one thing I am confident about is that the female teachers cared enough to look out for them. I am sure they will help them. They would not have come to talk to us if they did not see this injustice. One day it will change; I will change it when I am successful."

We all erupted in unison, "Yeah!!!" We sounded like revolutionaries.

Walking home, I got to thinking about what the girls had said. It reinforced my earlier view that all twelve of us were privileged and certainly the chosen few. I remembered one of the readings in religious classes, *"Many are called, but few are chosen."* How fitting. For some reason, I started seeing myself in a different light. I was not so wrong in thinking that I wanted a life different than my mother, grandmother, and aunts. Of course, I enjoyed learning about our traditions and cultures from them, but I had another idea.

I thought about my father. He had the advantage of being able to control everything he did while he delegated some things to others to do. I liked the place he had chosen for himself that gave him some independence. But I could not reconcile with how unfair it was for my mother to be denied the opportunity to go to school. The news that I was advancing to middle school assured me of a few more years to try and figure it out, an opportunity to find out why things were that way rather than being told how they were with no explanation. I was ready for this scary adventure, but I had eleven other girls with me to go through it. Also, teacher Catherine was going to be there, and we could always go to her. She would easily understand us because she already knew the kind of kids we were. At this point, I told myself, "No worries."

The last week of school was low-key. We were done with exams; we already knew what was happening the following year. There were a lot of impromptu activities assigned to us, including cleaning the church, its surroundings, and working in the school and church gardens. There were plenty of sports events, some friendly matches, and other competitions. The week went by very quickly, and we ended the school year looking forward to the Christmas holidays and a new year full of unknowns.

Preparing for Christmas was always a lot of fun. There was cleaning up, preparing local beer, cutting grass for the cows, washing, and ironing most of the clothes we owned. Often, we would go downtown to buy new clothes and shoes, mostly fabric that the local tailor used to make dresses for the girls and women, and shirts, trousers, and shorts for the boys and men. Men rarely bought new clothes for the occasion. They could get by with what they already had. For a child, this was the one time they hoped to get new clothes and eat a lot of good food. Families brewed lots of local beer, and sometimes slaughtered a cow, goat, or sheep. Families of little means might slaughter a chicken or buy meat from the butcher. Each family tried to make Christmas special, but those who could not afford it celebrated with neighbors. Invitations were not necessary, particularly for children. Anywhere you went you could eat, and adults got free adult drinks offered by the host. Children did not know much about Santa Claus, but there was an understanding in my family about what would happen if you were in Dad's black book. It could cost you a new Christmas outfit. The boys seemed to miss out more. My sister and I always had a new dress, shoes, and a shawl at Christmastime. If we were lucky, we might get two dresses instead of one made by the tailor from a cloth we chose from town or the local market. Having two dresses meant that we could wear one for Christmas and the other one on December 27, an important day in the village. This was the feast of St. John the Apostle, the patron saint of our church. It was treated like a day of obligation, with the priests offering multiple Masses like on Sunday. It was a public holiday for the village; nobody went to work. They went to church and then home to celebrate with good food and drinks. At our house, it was a major occasion because my father's name was John and people celebrated the feast of their namesake saint in a big way. We received a lot of visitors—many friends of my parents, the priests from the church, and neighbors. Because children could not drink the local brew, this was the one-time soda would be plentiful—lots of Pepsi, and Mirinda, no Coca-Cola or Fanta, their equivalence at that time. Some of the visitors gave us money, from one cent to ten cents. This was a lot of money, and we could buy lots of candy, which we much preferred over the drinks.

For me, Christmas shopping coincided with buying school supplies. I had received a letter from the school with a list of necessities. A week before Christmas, my father took me to town to pick up the items listed. Uniforms were not on the list. They were made at school and paid for in part by the school fees—two hundred and fifty Tanzanian shillings. The listed items included two sets of bedsheets, one towel, one pillow, and two pillowcases, a nightgown, one bar of soap for bathing (more could be bought at school), one natural fiber sponge, a whole stick of laundry soap (this was about sixteen inches long and would be cut into small pieces to last the entire six-month term), body lotion, flip-flops, and white canvas shoes, which were a part of the school uniform, to be worn every day. For the bedsheets and pillowcases, my father bought durable material that was cut and sewed by the village tailor. Interestingly, this was a special type of material from America. We called it *marikani*. The color was off-white with small grains of black that disappeared after a couple of washes. The material had a thick texture, making it popular for bedsheets, lining for window curtains (used as such in America too), mattress covers, and cushion covers. As bedsheets and pillowcases, the cloth was thoroughly washed and then dipped into a blue liquid solution to make it look brighter and crisper. Because of the thickness, it was necessary to iron to remove the wrinkles caused by wringing the water out since washing was done by hand. The final item bought was a medium-size suitcase to carry these items to school. These were not suitcases as we know them. Constructed from wood or metal, they were called boxes. At school, where they were stored was called the box room. The wooden ones were cheaper than the metal ones. Many students from wealthier families showed up with metal boxes. I did not know this until we arrived at school, and I saw bullies teasing girls with wooden suitcases. I do not know if my father knew this ahead of time and bought the metal one to spare me the inevitable teasing or whether it was a question of preference, and he had the money for it. It was brown with cream triangles, making it stand out.

After several hours of walking around town from one shop to another, gathering these items, we stopped at a restaurant. This was not my first time in a restaurant. I had made many other trips to town with my father to see the doctor. I knew what I wanted—a bowl of beef stew and bread. This was always a treat. I was like a kid at McDonald's when it came to going to a restaurant in town. This restaurant did something I have only seen at Olive Garden when it comes to soups. There was always an opportunity to have seconds for the same price. I saved the bread, eating the stew first and only starting on the bread after the soup was refilled. My father ordered the stew with rice. After eating, he paid the bill, and we left to go to a bread shop called Babu's to buy a loaf to take home. The family expected this every time my father went downtown. It was a treat since there was no shop in the village that sold bread during that period. Even then, bread was a delicacy, not something people ate regularly. He allowed me to get a bag of sweets to share with my siblings, a present from me from town. This was a big deal.

It was important to complete these preparations before Christmas because there was not enough time before school started. I packed my suitcase with the new items and selected a couple of church dresses and three every-day dresses to wear at school. We had been warned not to bring a lot of dresses because we did not need them. It was strategic. There would have been inequity between the girls because some came from families that did not have much. There were some who could not afford the school fees. Most parents had one income source, the sale of coffee. They could subsidize it by selling cereal like maize, beans, and millet, but these were extremely dependent on a good harvest in a particular year. My father and a few other dads were lucky, they had a salaried job, not much money compared to today, but a handsome subsidy.

My mother supervised the packing. She was home all the time because she had just delivered a new baby. Occasionally she talked to me about my being away from home for the first time. I could see her eyes welling with tears. I tried hard to avoid crying and would leave the room for a time. One time I came back and found her sobbing. I asked her why she was crying, but she would not answer. I started to worry, a feeling that stayed with me until I left for school. I noticed that every time she mentioned my leaving home for school to anyone who came to visit, she began to cry. The last week before I left was the worst; she cried every day. I started to notice that my leaving was affecting everyone in the family, including my father and my paternal grandmother.

A few days before I left for school, I visited close relatives like my maternal grandmother and both my paternal and maternal aunts. They did not show the same sadness, and I was encouraged. They gave me money to buy something when I got there and wished me well. One of my maternal aunts promised to visit me. Visiting my maternal grandmother was also part of my intention to see my sister before I left for school. She had spent most of her growing up with grandma, part of the culture where first-born girls are kept close to the paternal grandmother and the next girl is sent to the maternal grandmother. I was not sure if she would now be compelled to return home since I was leaving and there was no other girl to be of assistance to my mother.

Indeed, my sister Matilda was brought back home right after I left, and my brothers assumed more responsibilities like cleaning and helping with laundry.

— 3 —
LIFE IN BOARDING SCHOOL

Until my visit and stay in England many years after middle school, I did not know England had boarding schools. I thought they created boarding schools in the colonies only because they wanted to make it easier for the kids to focus on studying since being at home diverted their attention. Boarding school was also a good way to provide a level playing field for all the students because living conditions in the village varied depending on whether parents had a formal education, enough living quarters, or a conducive environment for structured study time. The middle school was geared to give us an experience that was close to that given to English children in boarding schools, a place for well-to-do and privileged kids. I realized that girls attending the boarding school with me were about to become the privileged few.

Finally, the departure day came after the monthlong Christmas recess. I had everything on the list. I washed, ironed, and packed everything neatly in my suitcase. The parents of the twelve girls going to Mandaka Middle School met and organized transportation to the school. Although it was possible to walk all the way, it would have been a tiring trip and a toll on our small, young bodies because we had our belongings to carry. We were grateful to our parents for being thoughtful and organizing a bus ride to school for the entire group. It was an easy task for them because one of the dads owned one of the village buses and another dad was the driver. The added advantage was that two of the dads were teachers at the school and friends of the bus owner. In the end, the ride was complimentary from the bus owner and driver and cost our parents nothing.

We were informed of the departure time and the pick-up locations. We were asked to be at that location at 11 a.m. Earlier that morning my father had wished me a safe trip and told me he left my pocket money with my mother. My mother showed me where she had tacked it, inside the pocket of one of my dresses that she had placed in the suitcase. My mother asked Anthony, one of my brothers, to help me carry my suitcase to the departure location. I started walking up the hill but turned around to say one final good-bye to my mother. I noticed, again, tears running down her cheeks. I could not bear it. I turned around and asked, "Would you rather I stay home?" She composed herself and responded while turning away from me, "Go, go, we will see you soon." I wiped away my tears and looked at Anthony who was also crying. What a disaster, I told myself. I turned around again, but this time my mother had moved farther away. I shouted back, "Okay Mama, see you soon," and left. Anthony and I walked halfway to the pick-up location without saying a word. Then I said to him, "You know there are visiting Sundays, the last Sunday of every month. Tell Mama to send two or all of you to visit and bring me something from home like sweet yams or sugarcane."

"Okay, I will tell her. We will come at the end of this month," he responded.

"Okay, I will be looking out for you. Make sure you tell Mama," I added.

"Sure, I will," he concluded.

We arrived at the junction, and I helped him with the suitcase that weighed heavily on his small head. We stood it on the grassy area and waited for the bus. Soon I heard the bus coming down the hill. It stopped. The driver got off, helped me with my suitcase, and I walked on. Anthony stood by the roadside and waved as the bus took off. Reality sunk in. This was it; the adventure had begun. I was on the way to my new home away from home, an experience I could not imagine. Two other girls were on the bus already. This helped diffuse the sadness of watching Anthony waving, and we immediately engaged in girl talk.

The driver made another stop to pick up more girls. This time four of them who lived close together were at the location. The driver helped them with their suitcases and continued to the next stop. Then we picked up one, then two more, and finally the last two. The bus was now rocking with preteens who were loud and excited about their journey, telling one another what they had in their suitcases and the snacks in their handbags. Then one girl suggested we should sing. We picked a song that extolled the school we were now leaving for good.

Ondokeni njiani, Ondokeni njiani	*Get off the road, get off the road*
Ondokeni njiani Iwa *ipite*	*Get off the road for Iwa (school) to pass.*

Ilikuwa ya kwanza kuanzisha masomo	It was in first place in providing education.
Ilikuwa ya kwanza kale hata leo	It was the first one before and today.
Shule ziko nyingi, ya kwanza ni Iwa	There are many schools, the top one is Iwa.
Kama husadiki, ngoja nikutume	If you do not believe, let me refer you.
Uliza ulaya we! hata Amerika we!	Check with Europe, even America.
Trala lalalala, Trala lalalala	Trala lalalala, Trala lalalala

The driver must have liked the song and the pounding on the chairs to get the sound of drumming. He honked as we passed people walking who stopped to watch the bus as it skirted around the windy road. We all waved at them as we sang. When we reached the junction of the highway and our main village road, we sang louder because there were more people waiting for transportation to different parts of the district. We made a left turn to go toward our new school. After a while, we decided to stop singing. We were tired and now out of the village. We just wanted to enjoy seeing the new landscape rolling hills through the windows as we passed by. By using the main road, the distance to travel was twenty miles from the village. After about four miles, the driver made a left turn heading toward the school.

We passed a few villages before we saw a large wire-mesh fenced piece of land. It looked like a tractor had just plowed. We drove for about eight hundred yards before the driver saw an opening that he drove through onto a feeder road, and we continued for about a quarter-mile. Then, we saw what looked like a mini town. The driver turned into the campus and pulled the bus near the first building we came to. We saw two-story buildings, one on the left and one on the right, a chapel in front of us, and three long buildings on the right. At the far north end, we saw four more smaller buildings. Later we realized there were more structures behind the north-end buildings. There were a lot of manicured lawns that we assumed were devoted to extramural or recreational/functional pastimes. We saw swings and seesaws. The view as we drove in was spectacular, with the pristine buildings, neatly manicured lawns, bougainvillea fences along the main entrance road, a huge vegetable garden inside the gate, healthy flower beds around the buildings, and sheer beauty. This was a designated private island for women only. No one was talking anymore, wondering what life was going to be like on this island.

The driver brought the bus to a stop near the building closest to the entrance. This was the classroom block on the ground floor with a dormitory on the first floor designated for the fifth graders. As soon as the bus stopped, a nun stepped out of a corner room of the building. It was Sister Agnes, the principal, and this was her office. She was dressed in a long-sleeved purple habit that dropped all the way to her feet and a white veil that covered her head, cheeks, and chin. The only skin we could see was her hands, the area around her eyes, nose, and mouth. The rest was fully covered. This outfit was accessorized with a woven belt that dropped in the front, and on the side hung three long rosaries clasped together and dropping down the length of her habit. All the girls were still sitting in the bus mesmerized. Sister Agnes came up the steps and said, "Hello little angels, welcome to Mandaka Middle School."

We had learned English since third grade and knew how to say thank you. We also understood when she said the word "come," signaling us to get off the bus. We did not realize that she did not speak a word of Swahili. Before all of us were out of the bus, an African teacher showed up. She was the female teacher on duty for that week. We had come a week ahead of the rest of the students for orientation. The teacher, Susan, welcomed us in Swahili. She spoke briefly with Sister Agnes, who then left for her office. Ms. Susan instructed us to gather our things and follow her. The driver handed each of us our suitcases, inspected his bus, and drove off waving.

We followed Ms. Susan, climbing the stairs next to the teachers' common room adjacent to Sister Agnes' office. Once we got upstairs, Ms. Susan pulled out a piece of paper from her pocket with the list of the students and bed assignments. The dorm was one large open space with a line of beds on both sides, separated by a small lamp stand. At one end was a big shelf with a curtain that dropped to the floor. This was the closet where we would hang our uniforms. On each bed were a mattress and a white bedspread. Ms. Susan started reading out our names and designating beds. She scattered us across the room leaving one or two beds between us. The empty beds were for other students joining us later. Some lived near the school and walked with relatives escorting them. Others came in on local buses and walked the rest of the way from the gate. We got there at two o'clock and by three o'clock all the students for the fifth grade were on campus.

Ms. Susan told us to unpack and taught us how to make our beds. She told us to unpack two dresses that we could wear for play or outside activities, underwear, soap, lotion, flip-flops, and school shoes. She had envelopes with our names on them and asked if we had any money with us. We put the money in the envelope. She informed us that we did not need the money until the end of the semester. It was going to be kept in a safety deposit box in Sister Agnes' office. She also announced that if we had any snacks, we should put them inside the suitcase. When all was done, she asked us to lock our suitcases and follow her with them in tow. She took us to a storage room where the suitcases were kept and reminded us that access would be granted once a week, on Saturday. Then Ms. Susan showed us where the outdoor bathroom, shower stalls, and clothesline were, and then proceeded to show us around campus. Because we were the first ones there, she allowed us to play on the swings and seesaws until four o'clock when everyone was on campus for orientation.

As we were playing, we saw other students arriving. At four o'clock, the bell rang, and we knew that was a signal we should assemble. We went back to the lawn in front of the classrooms where Ms. Susan and another teacher were waiting. We recognized her; it was Ms. Catherine. She smiled, and we smiled back. More students came from the dorm, the storeroom, and from a campus tour. Altogether, there were thirty-five of us. Ms. Susan greeted us again and introduced Ms. Catherine. She announced that our first orientation was to learn how to wash properly. That was a surprise, and we looked at each other. She told us Ms.

Catherine was going to lead that exercise and asked us to go to our dorm room and pick up a towel, soap, and a natural fiber sponge. Then Ms. Catherine led us to the front of the bathroom and shower stalls. She had prepared a big tublike basin, soap, a natural fiber sponge, and two towels, one the size of a hand towel and another a regular bath towel. She took a stool and sat in front of the big basin. We sat in three semi-circle rows, one behind another. There was a basket near her stool, which contained what she needed for the demonstration. She pulled out a baby-size doll, wearing a cute dress. She also pulled close to her a small bucket full of water. She proceeded to undress the doll except for the underpants. She asked us to name the parts she was pointing at—head, face, ears, neck, hands, chest, underarm, back, thighs, legs, and toes. When this was done, she stood the doll in the bucket of water, showing us how to wash each part we named to make sure the whole body was fully cleansed. She saw some of the girls giggling and stopped. She told them to stand up and then told the whole group that this was a real lesson on hygiene and not a joke. She warned the girls and continued with soaping and rinsing as needed. When this was done, she asked if anyone had a question. No one raised their hand. Then she announced, "Time for practicum. Who wants to volunteer to demonstrate what I just taught you?"

No one volunteered. I was sitting in the front row, right in front of the basin. Compared to all the other girls, I was the smallest in size. I was three feet, four inches. She asked me to volunteer. Because I knew her, and I was already her favorite from primary school, I did not want to disappoint her. I got up and took off my dress just like she did with the doll. I stepped into the basin. She told me I could kneel in it instead of standing. She gave me the soap and sponge and asked me to demonstrate while the others watched. Where I seemed to rush through, she told me to slow down or repeat and instructed the others to watch carefully. When I was done, she told everyone to clap for me. She took out the bath towel and helped me dry. She tied the towel around my small waist and helped me out of the tub. She then told me to put on my clean dress, and then go to one of the shower stalls to change my wet underpants. I came back and stood in front of her. She took out a bottle of lotion that read Yardley. It smelled good. She showed the students how to apply the lotion evenly on the body. When all was done, she asked the girls to line up in front of the ten shower stalls. One by one, they entered the shower stalls and washed as instructed. There were no doors to these stalls, but there was a long wall that ran the length of the stall area to prevent anyone from outside looking in. Ms. Catherine walked up and down the stalls to check on how everyone was doing. She allowed those who were done to head back to the dorm and apply lotion. I waited for the first bunch out and walked with them to the dorm. Though a strange first encounter, we were glad it was over, and we could look forward to other aspects of good living on this island. We had no idea what these nuns had in store for us. We were girls from the village; they came from a foreign land with no idea what a village was except from what the books they had read said.

We had a half-hour after the first orientation session. At six o'clock, the bell rang again. We left our dorm room for the assembly lawn. The teacher on duty, Ms. Susan, asked us to form one long line. We did and she led the procession to the chapel, which was right ahead of us. She led us, one by one, into the pews, about ten students in each one. She told us that was our permanent place in the chapel for the rest of the semester. Each time we came to the chapel we should take that same position, and when we get our prayer books, we could place them in the space under the bench. After that orientation in the chapel, Ms. Susan led the evening prayer, which took about ten minutes. These were a series of prayers from a booklet she handed us and one song. Afterward she led us out of the chapel to the dining room for dinner.

The dining room was a row of tables in a cafeteria style. Each table could hold sixteen people, eight on each side facing each other, and one place at the head of the table. There were two tables designated for us. There was one girl who had not arrived. Ms. Susan took the chair at the head of one of the tables and divided the number of girls equally at each table. The tables had already been set. There were no placemats, but each space had a plate, fork, and knife on opposite sides of the plate, and a spoon at the top end of the plate. There also were two big bowls with food, one with rice and the other with beans. On each side of the bowls was a big serving spoon. I was a bit familiar with this setup because of the influence my father had at his boarding school under the Holy Ghost Fathers. At home, we sat at a table for all our meals and each person served themself from a big bowl. We did not use a knife and a fork, instead, mostly a spoon. But whenever the pastor came for lunch on a Sunday, my father asked us to set knives and forks as well. Most of the time the pastor just used the spoon unless there was the need to cut a piece of meat. Even then, using hands was not considered strange.

In the dining room, the other girls looked a little baffled by this arrangement and waited for instructions. At this time, Ms. Susan had finished her part and a nun took over. Her name was Sister Perpetua. We stayed standing until Sister Perpetua finished blessing the food.

"Bless us, oh Lord, and these thy gifts, through Jesus Christ our Lord, Amen. You may be seated."

Sister Perpetua had a raised table in front of the two tables where we were seated, standing high enough to see all of us and for us to see her, although some had to turn their heads slightly. She called for our attention as she was about to do a demonstration. She said, "I am going to teach you how to eat with a fork and knife. Spoons are used for soup and dessert. For the rest of the meal, we use a fork and a knife. In the future, you will be taking turns in serving the food. Today, Ms. Susan and I will demonstrate for you. When you serve, you need to make sure everybody gets enough food, no more, no less. You serve small portions until everyone is served. If, at the end, there is food left in the bowl, you can do a second round. Unless someone says they do not want more food, everyone should get the same amount."

She came from her table and helped with the serving at the second table. Susan demonstrated at the table where she was seated. She had a plate and was going to eat with us. When everyone had food, Sister Perpetua started her instructions. She held a fork and knife high for everyone to see and said, "Hold your knife with your right hand and your fork with your left hand. Make sure the back of the fork is facing upward."

Ms. Susan repeated in Swahili, also demonstrating, because it was information overload in British English. Then Sister Perpetua continued: "Now immerse your fork in the food. With the help of your knife, load the back of the fork with small amounts of food and bring it up into your mouth. Watch how Ms. Susan is doing it."

We turned to look at Ms. Susan. She gathered a small amount of the food on the back side of the fork and brought it up to her mouth. She chewed her food and swallowed. She hardly moved her jaws much when chewing. We wondered how she could do that. She repeated the process while we watched closely, trying to understand this process, which seemed unnecessarily complicated. It was now our turn to try. Sister Perpetua walked around the table she was supervising while those sitting with Ms. Susan had the advantage of continuously watching her. We heard Sister Perpetua correct one of the girls. Some of the girls were having a hard time mastering this foreign way of eating, and since talking was not allowed, they suffered silently. It took longer to eat but practice makes perfect. In the end, everyone was done. Both Sister Perpetua and Ms. Susan congratulated us, and we all clapped. We stood up to say the end-of-meal prayer, led by Sister Perpetua.

"We thank you, Lord, for the food you gave us to nourish our bodies so we can serve you better. We ask this through Christ our Lord. Amen."

Sister Perpetua left the dining room while Ms. Susana remained to supervise the cleanup. Alas, we could now talk to one another, in Swahili, as we washed the dishes, dried, and arranged them neatly in a large cabinet in the corner of the dining room. We could not comment on the dining orientation because Ms. Susan was still with us. Finally, this was the end of the day. The day had been long, starting with traveling from our village to this new home away from home and plunging straight into orientation, which was intense, to say the least. We all deserved a good night's sleep.

Ms. Susan led us from the dining room to the dorm and supervised preparations for bed. We all headed to the bathroom where we took turns using the stalls. Once we were done, she led us upstairs to our dorm. She waited for everyone to get into bed, turned off the lights, and said good night. Per her instructions, no talking was allowed until after church the next morning. She must have stayed for a while until she was convinced that we were fast asleep.

It must have been five-thirty in the morning when we heard the chapel bell. Because our dorm was the closest to the chapel, it sounded like it was right outside the door. Two minutes later, the lights went on. Ms. Susan was standing in the doorway.

"Time to get up, girls, and head downstairs to wash. Take your toothbrushes, toothpaste, and a towel to dry your face."

This time she did not come with us. We knew how to navigate by now. It was amazing to get up so early and yet the lights around the school made it look like daytime. We hurried back to the dorm and changed into dresses and shoes. We followed Ms. Susan to church. The lights in the chapel were on, shining very brightly. Inside the chapel, on the right side, a group of about eight nuns knelt and prayed silently. They were finishing their morning prayers, which included chants and recitations. We later learned this was called "matins" or early morning prayers. As a group dressed in long purple habits, white veils that covered most of their faces, long woven belts, and three long rosaries hanging on their sides, it was a spectacular view. On the left side, there was another group, the African nuns. They were not participating in the same prayers as the white nuns but were praying quietly with their heads down. Ms. Susan led us to the pews right behind them. At the altar, two tall candles were already lit, and the table had been prepared for Mass. As soon as the white nuns finished their chants, exactly at 6 a.m., a priest emerged from a small room at the corner, known as the sacristy or vestry, to celebrate Mass. It took about a half-hour, and we were done. The priest had a parish nearby he was pastoring but also was assigned religious leadership at our school. This allowed us to have church services at the chapel without having to access these services at his local community.

After church, Ms. Susan led us to different chores around campus, mostly cleaning. A few of us carried wood from the woodshed to the kitchen and a group of four girls went to the kitchen to help put breakfast on the table. At 7:30 a.m., the bell rang, and we were instructed to go to the dining room for breakfast. It was a simple meal, porridge made with maize flour, water, and salt. The dining process was less complicated and so was the cleanup. By eight o'clock, we were on the assembly ground, ready for the day's activities.

Now that we knew where we were, it was time to be fitted with our new uniforms. At this school, the only time we could be out of uniform was when we had gardening and farming activities and on Sunday, after church. We only needed three outfits—school uniform, work dress, and Sunday after-church dress. If there was a special event, like an organized day trip or participation in a church event or a rally organized by the government, we had to wear our uniforms. Each student got two uniforms, a purple dress with a white collar and a white belt that was tied in the back. In addition, we had a white scarf to accessorize only when we went to church. Our white canvas shoes completed the look.

The nuns had a tailor on campus, one of the African nuns. Their order was the Sisters of Mount Kilimanjaro, and they were part of the diocese. But they had their own system, completely different from that of the Assumption nuns. They had their own house, across the street from the English nuns, but they cooked, washed, ironed, kept the chapel clean, supervised the vegetable gardening and general farming of corn, beans, and bananas, and looked after the chickens and pigs. For all these responsibilities, they were assigned a team of students to work with them where the activities directly affected the school. Students were not involved in anything to do with the English nuns, such as food and daily maintenance. The African nuns also worked in the school kitchen making all our meals. They served as the support system for the school and the English nuns. Most of the English nuns were teachers, nurses, counselors, and administrators. That is why, on the second day of school, we were sent to Sister Angelina to get measured for our uniforms.

The school fees were about $35 a year. I am not sure who was subsidizing the costs because this was a big operation. There were four classes, fifth to eighth grade, with thirty-five students in each class. The total annual income generated by the school fees was a mere $4,900, a lot of money in those days but still small for that kind of operation. Although most of the food was grown on campus, what we contributed as school fees (minus what we were taking out for uniforms) would not have been enough to pay the teachers' salaries, electricity, water, maintenance, and so forth. I do not think the African or the English nuns were paid salaries. From what we were told, they had taken a vow of poverty, and all they needed was a place to live and sustenance. Their service was geared toward doing God's will. It is without a doubt that the government of the time, under the British authority, and the local diocese were providing for the school and those who were taking care of us. At the age of eleven, we did not worry about money and who paid for what. We just wanted to be in school and learn something.

At the tailor, we were called into the shop, two at a time. The rest sat outside on the grass and waited their turn. When I was called in, I was curious how much material was going to be used for my two uniforms. Clearly, the concept of sharing to make what is available go a long way was not on my mind. I was fixated on the idea that my father paid school fees and whatever I got must justify what he had paid. I asked Sister Angelina how much material she was going to use for me since I was smaller than most of the girls. She told me I did not have to know that. All I needed to do was wait patiently to be measured and, if God loved me enough, have a cute uniform made. I was not amused. Sister Angelina realized she had just met her troublemaker. I insisted that I should get the rest of the material not used for my uniform because my father had paid for it. I asked her to look outside the window at a tall girl standing in line to come in. I claimed that she would have to take some of my material to get her uniform made and this was not fair. Sister Angelina put her hands on her hips and looked me straight in the eyes and asked if I wanted to get measured or not. When I realized she was a no-nonsense woman, I gave in to be measured while sulking. She finished measuring me and said, "There you go. It was not that hard, was it? I will make sure your dress is the cutest of all. Come in two days, and I will show you."

She pinched my cheek and said, "Smile, beautiful, smile. God will bless you, and you will pass your studies here with flying colors."

I figured she had won, and I had no choice but to be her friend. If she liked me that much, I was obligated to love her back. I gave up the fight and smiled. As I was walking out, she called out, "Do not forget to come and see me in two days."

I looked back and said I would not forget. Once measured, we could go to the playground until lunchtime. Getting on the swings and seesaw made me forget my encounter with Sister Angelina. But I did not forget to go back in two days as she had told me to. When I showed up at the door, she got up and embraced me like I was her child. She said, "Come and see. Isn't this cute?"

She had finished one of my uniform dresses, and I tried it on. There was a tall mirror in the room, and she made me stand in front of it and look. I did look cute, and I smiled. I turned around and threw my arms around her waist, which is how far I could reach with my height. She was happy for me. I took the dress off and gave it back to her. Then she said, "I have a gift for you."

She put her hand in her pocket and pulled out a rosary, then said: "you say this rosary every day and you will be successful. You will grow taller, and you will be very smart in class."

I admit that I wholeheartedly believed her. This was also an effective way for me not to forget her, a woman who showed me motherly love at my new home away from home.

The remaining days of the first week had a combination of activities with a good portion structured to allow us to bond as a class before the upper classes arrived. We started to develop the independence that we needed in this kind of environment, where one had to do what was right without being asked and to make decisions for ourselves, discerning what they liked and what they did not like. When we were leaving the village, we did not think we would be close to anybody other than our classmates from the fourth grade. We were surprised at how quickly that bond broke while we forged new friendships. My two best friends, Mary and Rita, came from different schools and villages. We looked out for each other the entire school year and made sure we gave each other the best presents on "feast days."

We did not celebrate birthdays, but our feast day, which is a church-designated day to honor one's namesake saint. It was customary for classmates and friends from other classes to join the celebrant and give them presents. The nuns made sure that the kitchen prepared tea at four o'clock with cookies (the British call them biscuits). The celebrant could invite up to ten friends to this tea, and while this group was celebrating, others were gardening or doing other assigned chores. It was during these times that friendships were kept or broken. If someone thought they were a friend and did not get an invitation, their feelings got hurt, and this could cause irreparable damage to the perceived friendship. Developing friendships across village boundaries helped to avoid the awkwardness when they were broken. It was not good to take the feud home during the holidays.

The weekend after the first week of school saw the full life of the sacred island, the school we were going to be at for four years. We were a unique group of girls from villages scattered around Mount Kilimanjaro. The sixth, seventh, and eighth graders started arriving on campus around eleven in the morning. We watched them as they arrived in groups, individually, accompanied by family members, or in local buses that made a detour to drop them closer to the school gate. They were all dressed in their school uniforms when they arrived. We now realized why we needed two uniforms. The uniform identified us with the school whether we were at or off campus.

The arriving students seemed like strangers to us as we did to them. We stared at them while they exchanged glances and walked on. They knew where they were going. They were veterans, not newcomers. The campus had four dormitories, and each building had a number on the side identifying the class. We started identifying the class levels of these new arrivals based on the buildings they entered. The teacher who was on duty this Saturday, Ms. Elizabeth, was moving around from one building to another, checking who had arrived. The box room was accessible because it was Saturday. We could go into the facility to get our Sunday dresses or snacks. On this day, the facility was open to allow the newcomers to store their belongings after taking out what they needed.

By one o'clock, all the girls had arrived. Ms. Elizabeth rang the big school bell announcing that it was time for lunch. There was no need for a supervisor to be in the dining room because we all knew what was expected. Ms. Elizabeth called the group to attention, said grace, and signaled us to start eating. It was dead quiet; there was no sound because talking was not permitted when eating. Halfway through, Ms. Elizabeth announced that the fifth graders would be clearing the dishes. Then she announced the weekend schedules:

"For dinner today, class six will set the table and clear the dinner dishes afterward. Class seven will fetch firewood for the kitchen before dinner today and tomorrow, and class eight will set the tables and clear the dishes for breakfast and lunch tomorrow. Finally, class seven will set the tables and clear the dishes for dinner tomorrow. Look for next week's schedules on the wall outside your classrooms."

Ms. Elizabeth finished the meal with a prayer and everyone except my class was free to go. By now we had mastered the after-meal cleanup, which included washing the dishes, wiping the dining room floor, and putting the dishes away. We had devised a system that had different girls doing different things for expediency. That afternoon, we did not have additional schedules. It was time for personal maintenance: laundry, ironing, and organizing our lockers for the school week starting Monday. Sister Angelina had finished all our new uniforms and handed them to us. Since they fit perfectly and did not need alterations, we could wash and iron them. They were already hanging neatly in the common closet at the end of the dorm room with our Sunday-best dresses.

The campus was now full of life. The girls were engaged in something, playing sports or just socializing. Before dinner, all the girls had to change into clean outfits. Sunday started with church at six-thirty in the morning, followed by breakfast. Because the term had not officially started, we all wore our Sunday dresses to church. After breakfast, we had an assembly where the teaching and maintenance staff came out to meet with us. The principal made announcements for the term and reminded us of the rules. These rules were posted on the school board in two columns. One column showed the rules and the other column the consequences for breaking them. One got a demerit for each rule broken. There was a sheet of ruled paper on the noticeboard outside each classroom with the names of each student in that class followed by the dates of the month. This was the demerit recording sheet. If one broke a role that required a demerit instead of a severe punishment, the teacher would put a mark next to their name. At the end of the month, the class teacher added the number of demerits and then decided on the punishment. Actions that warranted a demerit included being late for an event, not completing an extracurricular assignment, being messy—especially in the dormitory—talking in class or the dining room or being late to class after a break. One of the strange rules was speaking local languages, including Swahili, instead of English. We broke this rule frequently, not willfully, but accidentally, due to thinking in multiple languages. The fifth graders had the toughest time because it was a transition period from Swahili-only in primary school to English-only in middle school. English was now the medium of instruction except for the Swahili language class. This meant that ninety percent of the time, English was the language of communication on campus. To enhance proficiency, three English classes were offered each day, but Swahili was only once a week. Thus, as a rule, the teachers expected all students to speak English to them and to each other. To make up for our deficiencies in the English language, we created ideophones that helped move the conversation along. *Ideophones* are words that vividly evoke a sensory experience, like sound, movement, color, shape, or action.

The teachers worked hard to help us catch up. They often provided the needed word, then asked the student to repeat the word a couple of times. There was a sentence-building exercise to assure the teacher we had achieved competency. This proved to be the best way to learn new vocabulary words and to remember how to use them. Six months of soft immersion allowed us to keep up with the English nuns who did not speak Swahili and kept us from breaking the English-only rule. We could do targeted conversations, enough to allow them and us to understand each other.

The most interesting part was visiting the infirmary. Sister Veronica was the onsite nurse and the person to go to when a student had any discomfort like stomachache, headache, or had an accident on the playground. She also oversaw the convent house. We learned that her official title was Reverend Mother because all the English nuns answered to her. She made all the decisions pertaining to the other nuns' welfare as well as to the postulants (women who were training to join the Assumption nuns' congregation in England). The postulants also were part of the school staff. Some were teachers and others worked in the kitchens, animal house, garden, cereal farm, laundry or sewing room, and the infirmary. For Sister Veronica to understand us at the infirmary, we learned the key parts of the body to describe where the pain was located. Those who had gone to the infirmary shared their experience including the opening and closing of the conversation. Something like: "Good evening, Reverend Mother."

"Good evening my child, what is the matter?"

"I have a headache, Reverend Mother."

"When did it start?"

A successful conversation would end up with Sister Veronica dispensing a pill that the student swallowed at the infirmary. She also took their temperature to make sure the student was not running a fever, a sign of malaria. If the student had a fever, she would escort the girl to her dorm room, tuck her in bed, and inform the teacher on duty for that week. She would visit the student before evening prayer and early in the morning before church to take her temperature and dispense more medicine. If needed, especially in the case of malaria, Sister Veronica would get the driver to take her and the student to the nearest clinic for additional treatment. Her calmness and friendliness, coupled with her dedication to the students' welfare, made her the most likable person on campus. Some students would pretend to have a headache just to go to the infirmary to see Sister Veronica. She used to keep English mints in her pocket, and if she determined that the student was not actually sick, she would give her the candy and say it would keep her free from fever. We believed that, and when we found the candy was available in the shops, we nicknamed it the candy that kept fever away.

The first term at school was not easy. I did not think I would miss home, but the nuns were cognizant of the fact that we would be homesick. To alleviate this, they designated the last Sunday of each month for family members to visit. This was either a happy or a sad day depending on which families were able to send someone to visit. There were girls who never received a visitor while others had regular visits. Those who lived closer to the school were more likely to receive visitors than those who lived far away. Friendship saved the day as they shared their visitors and goodies brought from home. I did not get a visit the first Sunday and felt sad. When the clock struck three, I lost hope that anyone was coming to visit me. I never thought about the sacrifice they would have had to make to walk that distance. If I had, I would not have taken it so hard and felt so abandoned. I became physically sick and missed classes. The principal decided to send for my father who came promptly to the school. The idea was for my father to take me back home to a doctor and return when I felt better. My father arrived early in the morning, spoke with Sister Agnes, and then went to visit with Ms. Catherine before he called for me. He gave me news about everyone at home and a box of Marie Biscuits. This was a popular type of cookie, a gesture that made the visit special. He asked me if I wanted to go home, but I thought it was unnecessary. I was already feeling better. He told Sister Agnes that I did not want to go home because I was feeling fine. This was a perfect psychological fix, and I was back in class and normal again. The following visiting Sunday, my brother Venance and my sister came, the greatest gift from home. Thereafter, all went well, and I never got homesick the rest of the year.

Class started at eight o'clock, Monday through Friday, with a break at ten, lunch at noon, an hour recess after lunch, and finally ending at four in the afternoon. More time was allocated to English, three forty-five-minute classes (grammar, composition, and comprehension), every day. The other subjects—math, science, geography, and history were offered once daily. Swahili, religious studies, and domestic science (known as home economics in America) were offered once a week.

After-school activities varied. On Mondays and Fridays, we had sports, while Tuesdays, Wednesdays, and Thursdays were reserved for work in the gardens, farm, and the animal house (mostly pigs and chickens). There were not many options for girls' sports. Three were popular: volleyball, baseball, and netball (like handball or a modified version of basketball). The African teachers supervised these sports except baseball, which was taught by an American teacher, Mr. Philip. He came to the school each time we had sports, but we never found out where he lived or if he had another job. He was friends with the priest who taught religious studies and performed all the church services.

There was a rotation for other extracurricular activities, with one class doing one activity one week and switching the following week. The only exception was the farmwork, which was seasonal. All classes worked on the farm during sowing, weeding, and harvesting periods. Sometimes the school canceled classes to allow the speedy accomplishment of those tasks. With the gardens, they designated classes based on the task and the number of plots that needed attention. There was one English nun and one Tanzanian nun who supervised that, rarely the teaching staff except for Saturday gardening, which focused on flowerbeds. There was a permanent African nun who looked after the chickens and pigs and another who did laundry with help from two or three girls. Interestingly, the staff picked light-skinned girls more often than dark-skinned to help in the laundry room. We noticed this and thought they were receiving preferential treatment because of their light skin. It was not a surprise to us because in various parts of Africa, discrimination based on a girl's skin color was common and persists to this day. Most men preferred, and still do, light-skinned girls as potential wives. Sadly, this aspect has long been the major driver for skin-lightening products in cosmetic shops, a source of skin damage as well as the potential for skin cancer. At school, only the dark-skinned girls worked in the chicken coops and the pigs' den.

My petite body was an advantage. I did not work in the chicken coops or the pigs' den. But when I was in the sixth grade, I missed a highly sought activity—going on the truck to get cow manure for the gardens. This was popular because those who went on the trip came back with juicy mangoes to share with their friends. If you were not in their circle of friends, there was no mango. I wanted to be among the privileged few, but I knew there was no chance I would be chosen because of my size. They needed energetic girls who could carry buckets of manure to fill the school pickup truck. I started plotting strategies to overcome this dilemma. One night, I had an idea that included going to the truck ahead of the group and finding a place to hide inside the vehicle. I did not know if this would work or what the consequences would be if I were caught. I was determined to execute my plan to go on the trip for the mangoes, and nothing was going to stop me.

On the day of the plan, I was the first one out of class at four o'clock. I raced to my dorm room, changed my uniform to clothes I could wear when working in the garden, and set out to look for the truck. The driver had already pulled the truck near the classroom area. I realized he was not in the vehicle. Luck was on my side, so far. I ran to the truck and climbed from the passenger side to the bed of the truck. Originally, I thought that because it had a raised bed, I could sit in the corner, and no one would see me. As the girls came in, I could beg them not to say anything. What I had not realized was that I had more luck than what I had prayed for. In one corner was a pile of cloth that turned out to be the truck's canvas cover. The driver used it if it was raining. He also spread it over the manure to provide a comfortable seating area for the girls when they rode back. When I saw the canvas, I got another idea—to slip underneath it and stay as quiet as a church mouse. Because it was quite heavy, I unfolded one corner and then crawled beneath while pulling the canvas to cover me completely. A few minutes later, I heard the girls laughing and joking with one another as they climbed onto the truck. The driver, Brother Nicodemus from the Capuchin order, had lowered the tailgate for easy climbing and, one by one, all ten girls got on. Luckily, no one came to the corner where the canvas cover was and so I was safe. I could hear Brother Nicodemus locking the side brackets to secure the girls in the back of the truck. Then I heard him slam his door and start the engine. My heart was pounding whether from fear of being discovered or the joy that I had pulled it off. I waited under the canvas until I heard the driver apply the brakes. I came out from under the canvas with my finger on my lips, a signal to the girls not to reveal me to Brother Nicodemus. The girls started laughing, and I joined them. Brother Nicodemus lowered the tailgate, and the girls filed out. I was the last one off. I realized I was too small to jump from the height of the truck like the other girls. Brother Nicodemus looked at me and asked, "What are you doing here? How did you get into the truck without me seeing you?"

I did not say anything but looked at him sheepishly as if begging for mercy. He chuckled and extended his strong arms to help me off the truck. He retorted, "You are something else, you know that? I can report you to Sister Agnes, and you would be in a lot of trouble."

I lowered my eyes, then looked up at him. His heart must have melted. Then I said, "Please do not do that. I just wanted to get some mangoes. The big girls are very mean to us. They do not share their mangoes with those who do not come on this trip. They are very selfish. Please forgive me."

He looked at me and smiled. I looked like his little sister or something, desperate to do what the bigger girls were doing. Then he said, "Stay here and wait for me."

He unloaded the spades and buckets from the truck and took the girls to a pile of dry cow manure. Five of the girls were given buckets and five spades. He instructed them on how he wanted it done and how to spread it on the bed of the truck. They started to dig and fill the buckets, and then took the manure to the truck.

Once Brother Nicodemus was satisfied that they had mastered the process, he returned to the truck. He had a brown bag and told me to follow him. I was about to discover the mystery of the juicy mangoes. It was a huge tree laden with big mangoes. Some of the branches were quite low, but I still could not reach them even if I jumped up and down. Because I could not reach the mangoes, he picked enough to fill the bag he had brought from the truck. With a beaming smile, he handed me the bag and walked me back. We passed the girls who were working on filling the truck with manure. They looked at me as I walked past with my bag full of mangoes. Brother Nicodemus opened the passenger door of his truck and helped me onto the seat. He also handed me one ripe mango and said, "Go ahead, you can eat that one now." I smiled and dug my teeth into the mango, the juice running down my cheeks. He laughed, and I laughed too.

I was so happy that my mission was accomplished without harsh consequences. I did not think Brother Nicodemus was going to tell Sister Agnes about me sneaking out to go on the trip unauthorized. The big girls finished loading the truck with manure, and Brother Nicodemus gave them five minutes to pick mangoes. They all ran toward the sacred tree. I did not realize that on previous trips he had made a rule that no one could pick more than ten mangoes. He shouted as they ran toward the tree, "Remember the rule."

In five minutes, they were done, and Brother Nicodemus started the truck for our return journey. When we arrived on campus, he drove the truck to the designated location for the girls to unload the manure. He showed me a shortcut to my dorm behind the classrooms. I was grateful because I did not want to walk past the office of the principal, who would have easily guessed my crime. I would like to think Brother Nicodemus was aware of that and decided to be a partner in my crime and found a way to keep me out of trouble. I was grateful. I said thank you and ran into my dormitory clutching my bag of mangoes as tightly as I could. I also was looking around to make sure none of the sisters saw me and asked how I got the mangoes. Once in the dorm, I found a secure place for the fruit in my locker and could not wait to share it with my friends.

I did not realize that I had left an impression on Brother Nicodemus. I was now his little sister, and he looked out for me. Every time he went to get manure, he brought me a bag of mangoes that were delivered by one of the girls. I made sure I saw him in person to say thank you.

— 4 —

Just Being Kids: Surviving Middle School

Life in boarding school was always full of drama. There were friendship dramas, mischief, and simply innocent acts that could land a student in a lot of trouble. I had a big share in some of these dramas. A lot of them occurred on the playground. It was not uncommon to hear of a girl who was unkind to another girl because she was not friends with her. Because I was very talkative, I earned the position of class monitor. The typical duties of a class monitor included assisting the teacher in monitoring discipline in the classroom but could extend beyond that. I liked watching out for injustices and supporting those who felt wronged or bullied by others. Sometimes, this altruism landed me in trouble with the girls who were bigger than me or in the upper classes.

One day, in sixth grade, at the playground, I saw a girl being pushed off the swings. I approached the bully and told her off. I made her get off the swing and waited while the other girl used the swing to her satisfaction. We both moved to the seesaws, and I told the bully the swing was all hers. I thought we had settled the issue, but the bully carried it over to the next day. We were already on the seesaw when an older girl in the eighth grade approached me, held the seesaw to stop it, and ordered me off. I did and later found out she was the bully's cousin. I felt defeated and helpless because I could not take her on, a lesson for me for later about picking my battles.

Sixth grade was the toughest year for all of us. We had spent a year adapting to the new environment. We tried to be responsible but also wanted to be independent thinkers and decision-makers. But we lived in a new community full of rules, some good, some, we thought, ridiculous. We tried to understand the culture we were being groomed in, but we found it confusing because the English culture was superimposed on the African culture and a source of much of the trouble we got into from time to time. We often found ourselves at loggerheads. For example, the rule of silence from eight at night until after morning service, including at meals. We looked for ways to circumvent this rule, with dire consequences when caught. Our reasoning was that we just wanted to be happy and normal kids in school. We stuck together and tried to cover for one another when there was trouble. It was difficult for the teachers or the nuns to find out who the culprit was in presumed wrongdoing.

In the dormitories, the lights went out around eight p.m. The rule was that everybody should go to sleep in their own bed, and no talking was allowed. But as young girls do, we liked to sit on one of the beds and tell stories or gossip. To make sure we weren't caught, we would have one of the girls peep through the window curtains to make sure the teacher on duty had left the dorm area and had gone to her house, which also was on campus. We also knew that the nuns went to the chapel around nine for their evening prayers called compline or complin (end of day prayers). We took turns as sentinel. The guard would report when the duty teacher had left and when the nuns were in the chapel. We knew it took a half-hour for them to finish their prayers, and they would leave one by one for their residence, which was near our dormitory.

One day our watcher missed a step in her guard duties. She announced that the teacher was gone, and the nuns were in the chapel. We were now free to gather at one of the beds to talk. After half an hour, the guard was dispatched to the window to check the nuns. She saw them walking back to their residence, but she forgot to count to make sure all of them had left the chapel. She announced that all was clear, and we could resume talking. But, this day, Sister Agnes, the principal, had left the chapel through the side door and walked up the path along the row of dormitories toward her residence. She was on a fishing trip to see whether all the dorms were quiet, and everyone was sleeping. She cleared the first three dorms and then came to ours.

One thing we knew was that if any of the nuns were to come near the building, we would hear them because of the three rosaries they wore around their waist, which hung to the floor, and made a lot of noise as they walked. Sister Agnes had something else; she always walked around with her pitbull dog. That evening, she tricked us. She did not have her dog with her, and she held her rosaries in her hand to keep the noise down. She climbed the two stairs in front of the main dorm door and listened intently. She heard us talking. She had a pass key to each of the dorm's main doors. We think she must have inserted the key in the hole and waited for a few seconds before she turned it and pushed the door wide open. She switched the lights on, making it difficult for anyone to make it to their bed without her seeing who it was. The girls scrambled. Some did not have time to take off their slippers and climbed into bed still wearing them. Those who could not get to their beds thought they could hide from her by running into the closet. Sister Agnes walked to the closet, pulled the curtain open, and said, "Caught you beautifully, didn't I"?

She turned around and left the room. We could hear the click as she turned the key to lock us in. We knew we were in trouble but did not know how much.

In the morning, we went about our usual business until assembly time. After the prayer and announcements from the teacher on duty, Sister Agnes moved to the front of the assembly lawn. We could tell she was in a bad mood, and we were the cause. She started by praising the entire school for the good work we were doing and for keeping up good grades. Then she paused for about fifteen seconds and said, "But I am not too happy with the girls in class six. Last night I caught them out of their beds, huddled on one bed, and talking after the lights-out time. They have broken school rules; we are disappointed with them. To make matters worse, they tried to hide in the closet as if I would not recognize their beds were empty."

She paused again for a few seconds before continuing. "Now, these girls deserve to be punished as an example to all of you. School rules must be kept, no ifs or buts. Class six will work on the farm for the rest of the week as well as wash the dishes after each meal."

She left the stage and headed back to her office. The teacher on duty came back and dismissed grades five, seven, and eight and directed us to our dorm to change into work clothes. We did as she ordered and returned promptly to start our punishment. We could not be angry with the girl who was supposed to be on guard for us that night. We trusted that we were all in this as sisters, in support of each other. The lesson was learned, but we never let ourselves get caught again.

Other than the group mischief, I had a few of my own, both in and out of class. After my argument with Sister Angelina about my uniform, my next fight was with the librarian. This happened when I was in fifth grade. I had developed an interest in reading at the boarding school run by the Assumption Sisters from England. I liked to read a lot because I wanted to be good at English. I made trips to the library at least twice a week to exchange books. One day, I asked the librarian if I could check out two books because it was Friday, and I had the whole weekend when I could read both and finish them. The librarian flatly refused my request because that was against the library rules, one book per student at a given time. I knew that was because there was a finite number of books in the library. I calculated that some girls did not like to read and took the maximum two weeks to finish a book. I thought I could take those extra days a book stayed on the shelf and wasn't checked out. But this reasoning was not going to persuade the librarian to break the rules. She knew I was stubborn, and she could read my face that I disliked her answer. She was quick to let me know that she was not going to be swayed by my standing there and staring at her. She advised that I leave and come back Monday.

I was not happy about this, and I told her she was being mean to a little girl who liked to read. She looked away and ignored what I had said. I quickly realized she was not going to be like Sister Angelina, so I left the library with my one allowed book. I went to visit Sister Angelina to tell her about my disappointment at the library. She was sympathetic and told me to come back in the afternoon. I did, and she had found another book for me, one about the lives of the saints arranged in alphabetical order. Unlike the one my father had; it was written in English. I asked her when she wanted it back, and she said I could keep it until the end of the term. I was happy and forgot about the encounter with the librarian. From this, I found a new mission, telling all the girls about the life of the saints associated with their names.

In class, I didn't have many troubles or disappointments, but one that left an indelible mark on me was annoying my geography teacher. I had one problem—I was not very patient and could not sit still if bored. My parents, especially my father, thought the solution was to keep me busy, but my teachers' solution was punishment. My boredom in class stemmed from a habit I had developed as a child in a household where being prepared was enforced by my father. When he prepared for his next day's classes, he made us prepare too. As a result, we were always ahead of the class on the new material being covered by the teacher. Dad did not entertain idleness. If he asked, "What are you doing," you could never say "nothing" because it gave him a reason to find something for you to do. The easiest thing was to look busy to avoid that question. It was routine for us to sit with my father after dinner to look at the math problems for the next day and tackle them ahead of time. If there was a problem, Dad stepped in and helped by posing leading questions to allow us to find the answer ourselves. This made it easier for us to understand what was being taught and allowed the teacher to assign us to help a struggling student in class. I always had someone to assist in class when I was in primary school, partly to keep me busy.

I started being a problem in middle school because I retained the habit of being overprepared for my next classes and became bored when I was done with my work and did not have something else to do. Often, I became restless when the teacher took too long to explain a concept, and I would volunteer to simplify it. That is how I started getting into trouble with my geography teacher, Ms. Theresa. When she wrote a class assignment on the chalkboard, I raced through it and finished it just after she was done writing it up.

One day I decided to be creative by taking my handicraft work to Ms. Theresa's class. I hid it in my desk drawer but pulled it out as soon as I finished her assignment. Ms. Theresa saw me and told me to put it away. I told her that I had finished my work, and I was bored. She asked me to stand up and stay standing until further notice. I did, but my patience did not last long. I put my hand up, but she tried to ignore me. I started snapping my fingers to get her attention. She became more annoyed with me and asked what I wanted. I told her that I saw a fly zooming around the classroom, and I would like to try and kill it. At this point, she had had enough of me because a week before I had made the class laugh by making funny faces while she was writing on the board. She kicked me out of class for the rest of the period. Instead of going to the dormitory or library or standing outside the door, I went around the building to a spot near our classroom window where everyone could see me except Ms. Theresa. I made funny faces, and the students laughed. Because she saw the students looking out the window, she came over and saw me squatting under it. She told me to come back to class. There were five minutes left before it was over, and she knew we would be out soon.

Ms. Theresa gave us our homework and as she was leaving class, she said to me: "I do not know what to do with you. You are smart but a lot of trouble."

I did not misbehave for a while making Ms. Theresa believe that I had calmed down. It was a surprise to her when I started again. Asking to kill flies triggered her memory of my past mischief. She continued to ignore me, leaving me standing until the end of class. When everyone left for break, she asked me to remain sitting at my desk. She also asked two other students to remain in class until she returned. She left briefly, came back, and told the two girls to move my desk to the fifth-grade classroom, next door. She had spoken with the teacher and explained that I was going to be in her class for a week as a form of punishment. After the other students moved my desk, the teacher disclosed my punishment. I did not argue with her this time, and after the break, I moved to the fifth-grade classroom. Because I did not have to learn anything there, I decided to use the time to read. I went to the library and found a book called "The Guinness Book of World Records." The title impressed me, and I decided to check it out.

Because my punishment started on a Wednesday, I was required to remain in that class until the following Wednesday. But the plans changed midweek when we had a surprise visit by the district education officer, Ms. Bryson.

Although our school was considered privately owned by the Assumption Sisters, they were bound by the mandates established by the Board of Education. The board monitored the quality of the curriculum as well as teaching practices. Ms. Bryson, from England, was required to visit all the schools in the region once each term to conduct this evaluation. She never announced her visits beforehand to ensure that the school did not put on a show for her. Upon her arrival, she did not wait to be received by the school's administrative staff. She just picked a class to inspect. Afterward, she went to the teacher's office to look at their teaching plans and lesson notes. She also inspected test results and other materials of her choice that were kept in the principal's office.

Unexpectedly, on Friday morning, two days after I moved to the fifth-grade class, Ms. Bryson arrived at our school. As soon as she parked her Land Rover, she headed to the fifth-grade classroom. My desk was by the door, so she saw me before she entered. The students stood up and greeted her in English, "Good morning, madam."

"Good morning girls. Please be seated," she responded.

"Thank you, madam, welcome to class five," they replied.

Ms. Bryson went to the front of the class, exchanged greetings with the teacher, and then took a chair near the window. She told the instructor to continue teaching. She sat there for about ten minutes observing how the teacher was engaging with the students. The teacher assigned the students some group work while she wrote English sentences on the board. Ms. Bryson took the lesson plan notebook and flipped through it. She then got up and walked around the classroom to see what the students were doing. She spent about fifteen minutes before leaving. Before doing so, she randomly picked five exercise books from the students. She wanted to check the notes the students wrote and how well they did the assigned homework.

Ms. Bryson noticed that I was not doing what everyone else was. I was busy reading my book. My desk also was out of line, clearly showing that it was an extra desk in the room. Each of the classrooms had desks for thirty-five students.

Mine was extra, a violation of the rules. She did not ask the teacher about me, but went straight to Sister Agnes. I was not aware that the principal did not know I was being punished. She was taken by surprise when Ms. Bryson asked her about the extra student in the fifth-grade class. Both Ms. Bryson and Sister Agnes returned to the class, stood by the door, and summoned me to join them. I could see that Sister Agnes was furious and asked me what I was doing in that class. I told her I was being punished by my geography teacher. I was surprised that Ms. Bryson would address Sister Agnes in my presence. She told her this was unacceptable and that there were better ways of handling students. They turned around and went back to the office while I returned to the classroom.

Ms. Bryson completed her visit and left before lunchtime. After lunch, Sister Agnes ordered my desk back to the sixth-grade classroom, and that was the end of my punishment. I was happy to return and that I did not have to miss any more classes that week. The teacher did not look happy. I felt remorseful and made a promise to myself not to give her any more trouble in class. I realized that I had caused trouble to both Sister Agnes and Ms. Theresa, but I did not know what to do about it. Sister Agnes was not easily approachable, so I decided to go to Ms. Theresa and apologize. I was glad when she accepted my apology and asked me whether I would like to help her in class whenever I finished my work early. I agreed, and we became friends.

Another mishap stemmed from trying to be creative in an unconventional way. The thinking that landed me in trouble was similar to what prompted the argument with Sister Angelina regarding leftover cloth from the making of my uniform. I thought that I did not need a mattress that filled my bed when I was only four feet, three inches tall. My rationale was that when I slept in the bed, I had two feet of mattress space unused. I decided to make good use of this extra material on the day we were told to refurbish our mattresses by filling them with dry maize (corn) husks.

The maize husks made the perfect filling for these mattresses instead of expensive cotton wool or kapok fibers. The maize husks also were a source for making creative crafts like dolls, traditional dance skirts, and decorations. Thus, when we harvested the maize, we saved the husks for creative activities. The cobs were used like wood for cooking and for hot charcoal and like irons for pressing our uniforms and other clothes. Every student looked forward to the mattress-filling event each year during the second school term (June-December). The same people who made our uniforms made our mattresses and pillowcases. It was the students' responsibility to fill them with the maize husks and hand-sew the opening. We did most of our needlework by hand, a skill we learned in home economics class. We were taught how to sew, knit, crochet, and embroider. Students in the upper classes filled the new mattresses as a welcome gesture. Thereafter, each student made their own.

Filling the mattresses and pillowcases was a lot of work. First, we removed the old husks and washed the mattress pockets. Then, we thinly shredded the corn husks. This process usually started early Saturday morning and by lunchtime half of the work was done. In the afternoon, the process of filling the mattress pockets with the new husks started and continued until it was done. Then we took them to the dorm and neatly made the beds. They looked ridiculously funny because some girls had filled them so high that it was hard to climb into bed. This did not last long, about a week, when the husks softened and leveled. Some girls fell off their mattresses a couple of times before the husks were soft enough to stabilize.

It was while filling my mattress that I got the creative idea that landed me in trouble. I took a ruler, measured the excess material, took a pair of scissors, and cut it. I sewed it neatly before putting it on my bed. After making my bed, I pulled the bedspread to cover the bare metal that was not covered by the mattress. I am not sure why the other girls did not say anything to me because my bed looked distinctly different than the others in the room. I did not think I would get into trouble for this; I was elated that I had had a cool idea. I was also happy to have extra material to make dresses for my dolls. I had two dolls that Sister Thomas, one of the nuns who I liked, had given me as a present when she came back from a trip to England. She was my English teacher, and I was the class monitor. I always helped carry her books back to the office after class. She went to England for a vacation, and before she left, I asked her to bring me back something nice. She kept her promise and brought back two elegant dolls. These dolls sat on the lamp stand next to my bed during the night but during the day I placed them on my bed lying side by side. The only problem I had with these dolls was that they had only one outfit each. If I wanted to wash them, they stayed naked until the clothes dried. This bothered me a lot. After cutting my mattress, I found a solution to this problem. I spent most of Sunday afternoon hand-sewing a second set of outfits, a couple of tiny skirts and blouses. I was thrilled by my creation, but trouble was looming.

Early Monday morning, I completed my morning chores and went to class. After the first period, Sister Agnes came to our class and summoned me to follow her. She did not say a word until we arrived in the dorm. She stopped in front of my bed, turned to me, and asked, "What is this, may I ask?"

"My bed," I answered.

"What about your bed?" she asked.

I knew what she was getting at, but I played dumb. I did not answer her question. She looked at me again and asked the same question. One thing about Sister Agnes was that she had a distinct feature on her face. One long hair on her left upper lip. When she was perturbed, we knew right away because this piece of hair would stand up like a cat's whiskers. I held my silence.

"What happened to your mattress?" she asked.

"I cut it," I answered.

"Why, young lady?"

I raised my head to look at her and began my explanation. We had been instructed to address her as Mother.

"Well, Mother Agnes, to start with, this mattress was too long for me. When I slept on it, I only occupied the portion that you see on my bed now. The rest was a total waste. I had good use for the extra cloth."

Before she said anything else, I moved to the side table near my bed, opened the door and pulled out the tiny outfits.

"Look, Mother, I made these for my dolls."

She took them from me, gave a look, and started walking toward the door. She turned and said, "Go back to class."

I followed her out of the door. She turned right heading to their convent house, and I turned left to go back to class. I did not hear anything said to me the rest of the day. I did not tell any of the girls what happened in the hope that my excuse was well taken. Two days later, shortly after the ten o'clock break, Sister Agnes came to my class and took me out again. She asked me to follow her to her office. I had the shock of my life when I got there. Sitting on a chair was my father. I said hello and moved to the other corner of the room and stood there facing him. On Sister Agnes' desk, I could see my work of art, the clothes I had designed for my dolls. Sister Agnes started talking. "Yes, Mr. John, your daughter has been very naughty. She cut her mattress and look at what she made."

My father took a good look at what I had made. I thought he was going to blow up at me. He was not smiling, but he had a happy face as if admiring my work. He then said, "Sister, I think she has too much time on her hands. Keep her busy. Give her more work."

I thought Sister Agnes was going to explode. Her lip hair stood up again, and her cheeks looked puffed up and red. Sister Agnes paused for about fifteen seconds and then said, "Well, Mr. John, you can take your daughter home with you, and she can come back after a week."

My father did not say a word. Sister Agnes told me to change my uniform and wear a home dress and come back to meet my father at the office. I held my tears until I was out of the office. When I returned, my father was outside sitting on his motorbike. I climbed on behind him, and he drove off campus. He did not say a word all the way home. When my mother heard the motorbike, she came out of the kitchen. She looked baffled when she saw me.

"What happened?" she asked.

All my father said was that I was going to be home for a week. He turned his motorbike around and left. I told my mother the whole story and she seemed sympathetic. We both thought my father had gone back to school only to be proven wrong. After an hour, we heard the motorbike again. I ran into my room and waited there. I heard my father asking for me; my mother called me out. My father was holding a big bundle of fabric and a bag with needles, thread, and a pair of scissors. He looked at me and said, "Here, I want you to sew until sewing is out of your system. I want you to make me bedsheets, window curtains, tablecloths, and if you have cloth left, make me handkerchiefs."

I took the supplies from him and headed back into the house. I heard his bike leaving. I placed everything on the dining table and looked at it, wondering how I was going to meet his order, sewing by hand. Come to think of it, this was an instructive punishment in place of yelling and fussing about what I had done.

When my father came back from school, he did not say anything more about the whole incident. It seemed all normal family life except that I was home when I was supposed to be at school. Both my parents were accommodating and helped me get through the week at home. I worked hard on my father's tailoring orders and by the end of the week, I had everything done—two sets of bedsheets, three tablecloths, and curtains for his bedroom and the dining room. With the remaining pieces that were too small, I made him the handkerchiefs he wanted. I even made more outfits for my dolls, but I did not disclose that to him. He examined everything and praised me for a job well done. Then he became serious and said, "Look, when I sent you to Mandaka, I wanted you to be a good student and pass your exams. I did not send you there to get into trouble. Make this the first and last incident I ever hear about."

I looked up and assured him I would do that. I prepared for my return to school. He took me on his motorbike again, but when we arrived at school, he did not wait to speak to anyone after dropping me off. My friends were happy to see me back and helped me catch up with notes from the different classes. I studied hard because there was a midterm coming up. I was sure I would not do well but my friends' efforts made it possible. When the results came out, I was surprised that I ranked third in the class. Two of my friends ranked first and second. We decided we would celebrate the following Sunday when the storage room opened, and we could access our snacks.

After all this, one would think that there would never be any more escapades from me. But I was in a group of girls who were sent home to bring our parents to campus again. This time it was an innocent act, something we did at home, and we thought it was fine only to find out that it was not the case with these English nuns. A group of us had been sent to the farm to cut maize for dinner. After cutting what was enough, we decided to chew on the maize stalk because it tasted like sugarcane. We were all used to doing that at home during the maize season, but it was unthinkable at school. While we were having this maize-stalk eating party, someone saw us and reported us to the authorities. We did not know about it until the next day during assembly when Sister Agnes summoned us to her office. She told us what our offense was and sent us home to bring our parents. I was devastated. I did not know what my father was going to do with me after promising not to get into trouble again. I had no choice but to go home.

When I got home, my father was still at school, but my mother was at home. She was shocked to see me, but when I told her what had happened, she appeared sympathetic but unsure of how my father would react. She gave me something to eat, and I changed into home clothes so I could help her with housework. When my father came back in the evening, my mother did not tell him immediately that I was home. After a little while, she did. My father asked her where I was, and she called me from the kitchen. I greeted my father and was surprised he responded calmly. He asked me what had happened, and I told him the story. He chuckled and then kept quiet for a few minutes. I did not think this was funny, but he told my mother later that the English nuns were ridiculous. My father went about his planned activities for the evening and did not mention it at dinner. He told my mother to let me know that we were leaving early in the morning to go back to school.

Early in the morning, we started our journey back to school to face Sister Agnes' wrath. My father wanted to get it over with early, return home, and get to school so he would not disrupt his teaching schedule. We got back to school just after the assembly. He parked his motorbike, and we headed to Sister Agnes' office. When she saw him, she stood up, and they greeted each other. She told him what the group had done, but my father did not comment. He looked straight at her without any reaction, as if asking, what do you want me to do? Then Sister Agnes said, "Mr. John, do you understand what these girls did? It is bad."

My father did not answer right away, and Sister Agnes waited patiently. Then my father asked her, "How much do I need to pay for the damage?"

Sister Agnes responded, "Surely, Mr. John, you do not understand."

My father replied, "Sister, I do. I am also a teacher and kids do stupid things." Sister Agnes paused and then said,

"This time she is not going home with you, but if she does something else, she will be sent home for a week."

My father did not say anything, turned around, and headed for the door leaving me in Sister Agnes' office. I heard his motorbike leaving and he was gone. Sister Agnes looked at me and told me to go to my dorm, change into my uniform, and go to class. I breathed a big sigh of relief and ran out of her office. One by one, all the other girls returned, and by the end of the day, everyone was back on campus. In the evening, we shared our stories about the encounter between Sister Agnes and our parents, and as kids, we just laughed it off. We thought it was much ado about nothing. But rules are rules, and we broke one. This was the last time I had a run-in with Sister Agnes. I made an extra effort not to cause my father any more grief over my mischievous deeds. When I went home for the break, my father mentioned the incident and asked me again not to get into trouble. He said, "If you can finish your time at Mandaka without having me called to the school again, I will give you a good gift."

This was a great incentive, and I made sure I did not jeopardize this promise. I succeeded, and at graduation, he bought me a watch. I was the first and only child he ever bought a watch, and I treasured it. Because it could not be worn with our uniforms, I only wore it on special days and when I came home for holiday from high school. It was the only watch I had until college when I was able to buy another one on my own.

HIGH SCHOOL

We did not have to worry about what we were going to do after eighth-grade graduation if we maintained good grades.

The nuns had it all figured out for us, creating a long vision and foundation for our future. They decided to expand their mission to educate girls by opening a high school in the region. When we arrived at Mandaka, the first group was entering eighth grade. A year later, the high school was completed and could accept the first group of graduates from Mandaka. The goal was to provide continuing education for as many girls from the middle school as possible. At that time, there were two Catholic high schools for girls in the entire country. Missionaries of other denominations also were scrambling to create educational opportunities for women. Before independence in 1961, the different religious groups operated individually. As kids, we were not encouraged to associate with someone of a different religion. After independence, the central government decentralized the education sector, eliminated the monopoly exercised by religious organizations, and eliminated segregation based on religious affiliations. The nuns also were following the changes that came with independence. They were aware of the government's intention to remove the single-sex middle schools, allowing girls to be in coed middle schools in their respective villages. The nuns saw a need to focus on high school and discontinue the middle school. Thus, the new high school was timely, a move to consolidate their resources that would have been split between middle and high school and a continuing education guarantee for the girls who were currently in their care. Some of the teachers moved to the high school while others remained to start phasing out the middle school.

The high school was named Assumpta Secondary School, derived from the name of the nuns' order. They still had autonomy and a bigger say on who could come to the school but were required by the government to ensure equal opportunity regardless of religious affiliations. Nevertheless, the girls from the middle school were given priority if they showed promise. Grades were important and the competition was intense, more so two years after my class finished eighth grade because the government mandate became effective then. The high school started with single-stream classes for the first three years of operation. Thereafter, the government required double streams for each class (two classes of the same level) to allow more students to come to the school.

Once established, there was fierce competition to be accepted as a student at Assumpta Secondary School. The buildings were spacious and well-equipped with science labs. The dormitories were different from the middle school. Instead of a large hall with a row of beds on each side, there were partitions to make cubicles with two beds, a closet, and a working desk with drawers for storage in each cubicle. The bathrooms were in the same building and accessible through a door at the end of the row of cubicles and continued to a laundry area where the students could wash and dry their clothes. This was a major upgrade from the middle school where the bathrooms were outdoors and shared by all the students.

Assumpta Secondary School was the place aspired to by middle-school girls in the region and daughters of rich families and high government officials. The Tanzanian president's daughter was sent to Assumpta. I had the opportunity to know her well because of a buddy system established by the school administrators, which paired up newcomers with veteran students. The pairing was called "the family tree." Each student was assigned a mentor who was referred to as "mother." The mother's responsibilities included providing guidance to the new student, giving advice on social and educational matters, and protecting the new student's interests. In the following year, when the new student became a mother to another student, their mother became a "grandmother." Then, the system added another layer, "great-grandmother." This structure provided stability for the girls, creating a home away from home and a surrogate family while at school. As a family, they met occasionally, especially on Sundays, organized birthday and patron saint celebrations, and exchanged gifts at religious and other occasions, like graduation.

My school family was privileged because I was in the weekend kitchen support group. On the weekends, the cooks did not come to campus. It was an opportunity for those who wanted to cook to be the chefs for the weekend. I volunteered every weekend, mostly because there was payback. The privileges included an opportunity to come to the kitchen and sample some of the food before it was served. Sunday dinner was soup and dinner rolls. If we had leftovers, it was the chef's prerogative to decide on the recipients. My "family" was often a priority before friends. With the extra cash earned, I could afford expensive presents for members of my school family. My sister, who joined me at the school in my third year, also was included in my preferred group.

My connection to the president's daughter was my mentee who became her mentor. For me it was coincidental, but my mentee was a designated match because of who she was. She came from a well-known family. Her sister, a Tanzanian, was married to the Minster of Health, a British national who had been naturalized in Tanzania. It was interesting when family members of these two students came to the school because I was asked to meet with them. They often asked about their daughters' progress at school. They brought me thank-you gifts, mostly cookies, candy, or decorative handkerchiefs. The president stopped at the school whenever he was nearby. His appearance was dramatic. We would be in the middle of class activities and then a big Land Rover would appear at the gate with soldiers and security officers. After exchanging words with the guard, they would drive straight to the principal's house. A message was sent to the teacher on duty to alert all other teachers to gather students on the assembly grounds to wait for the president's arrival. The visit was usually brief, and although the nuns tried to show off the president's daughter, the president made it seem like his intention was to see all of us. He was a teacher before he became president and took pleasure in joking with the students, picking on anyone, and making all of us roar with laughter. His stopover was never more than fifteen minutes, five with the nuns and ten with the students. We considered ourselves privileged to have an audience with the president of the country without a lot of security protocols.

The four years at high school were still formative years for the girls, a continuation of the previous four in middle school. As we approached the end of twelfth grade, we became worried about two things. One was the ordinary-level examination (officially known as the O-level exam) that was required to determine who could go to preparatory school (two years of study to prepare for college). The preparatory school was popularly known as Form Five (year one) and Form Six (year two), a continuation of Form Four (twelfth grade). The second concern was the possibility of returning to life in the village after eight years of sheltered life at boarding school.

The scary part of the O-level academic evaluation was that it was set in England and was similar to what students of the same age and level of education in the British colonies took. The exam was based on a curriculum designed by the British Examination Board in England, and teachers were expected to ensure that the students learned all that the British thought was necessary. Little was relatable to our environment or local conditions. The teachers taught the books assigned, and we memorized the facts. Our responsibility was to reproduce what we learned from the books as directed by the teachers regardless of the usefulness to our future lives. The teachers studied the patterns of the exams and discovered that some of the questions were repeated in subsequent years. To prepare for the exam, it made sense for the teachers and the students to use past exam questions. Eighty percent of the time, the teachers were right; past questions showed up in subsequent years.

The exams were administered in November for about three weeks, one exam each day including science and home economics practicums. The exam papers were shipped back to England to be graded, and after two months, the results were sent back to the Ministry of Education. After the exams, some girls suggested that we should start a novena to ask God to strike the ship carrying the papers and capsize it in the middle of the Indian Ocean. I do not think we knew what that would buy us, but it was just a thought. It never happened. Once we finished taking the exams, we left the school for the Christmas holiday with heavy hearts because we did not know what lay ahead in the new year.

Many of our teachers told us before we left that we did not have to worry about anything. We had worked hard for four years and that should count for something. But we were worried about life in the village after four years at boarding school if we did not pass. We realized that we were about to be unleashed into the world where the protective coat provided by the care of these nuns would be stripped off our backs. There were no more guarantees. None of us knew what would happen without the structured life we had enjoyed for eight years. I started thinking about how mischievous I had been in middle school, how much I had grown to be responsible for my actions. Yet, I was not confident that I was ready to make decisions for myself. I needed more time and more tools. I thought about my geography teacher in middle school and wished I could go back and say I was sorry again. All of us needed to thank the nuns who left their comfortable homes in England to prepare us for adulthood. All of these good thoughts did not take away the insecurity; we felt naked in a hostile world. The security provided by the four walls that were enhanced by a strong fence and a gate, which had a 24-hour guard, was no longer available to us.

We celebrated the end of the exam period even though we were worried about the outcome. The remaining weeks before the holidays were used for cleaning the school compound and packing our belongings, ready to go home. It was customary to have an end-of-school party to say goodbye to the graduating students. The administration made it special with rare snacks, a good meal, and an evening dance party. It was a special evening when we did not have to go to bed early. They allowed us to dance until midnight.

… 5 …

New Beginnings

It was December 29 1967, after a morning service at church, when the priest at our church told me there was a letter that came into their mailbox. Most people in the village used the parish address for contact. As a teacher, my father was very well known at the parish. They also knew us because my father invited them to our house many Sundays for lunch. This relationship was the reason he made it special by telling me to go to the office and pick up the letter. Because it was addressed to my father, I could not open it, but took it straight to him. It looked like it came from Assumpta Secondary School. I began to worry, not knowing what the contents might be. I thought it could be about my sister since we went to the same school, and she was returning in January. The school administration usually sent reports home to the parents that showed the progress of their child for the entire year. I knew I was not going to get a report because I had just graduated. My sister and I rushed home and handed my father the letter. We did not hear from him about it for the rest of the day. My father was never rushed about anything. He took his time. His silence led me to assume that this letter might concern my sister and, if the grades were bad, we would all hear about it after dinner. But if the letter concerned me, I would endure a longer lecture about how my road to a bright future had come to a sudden dead end because I did not do well enough to go to college prep school. I did not think that my father would get any pleasure giving this lecture. It would have been very painful because of the hope he had put on my success. I thought about how I would feel because this would be an outright betrayal of the trust and hard work he put in to help me excel. My stomach was in knots all day, but I hoped that the novenas my friends made all year would not have gone unheeded.

At dinner, my father did not show anything one way or the other that would have given away his mood. He made his usual small talk, teasing or picking on any one of the boys for fun. Sometimes we would laugh, and sometimes we did not because we could not figure out where he was going with a particular joke. If he laughed, we also laughed. It was easier when he talked about general things with my mother, and we listened in. After dinner, before we got up to clear the table, he looked at me and said, "I got a letter from your school."

He paused for a few seconds and then continued. "You have been selected to go to preparatory school, Korogwe Secondary School, for your two years before you can go to college."

He was now smiling, and my brothers and sister were smiling. My mother said, "good job," and added that her prayers had been answered. I was still trying to recover from the cloud of uncertainty and fear I had been experiencing all day. My father added that this was a good example and expected my sister to follow in my footsteps. He led the prayer after dinner, and we embarked on clearing the table. When we got outside to wash the dishes, my sister and I held hands and jumped up and down with joy and jubilation. My light was on again. I had bought myself two more years of a sheltered life. I went to bed relieved and looking forward to a time when I would catch up with my friends to find out if they too had learned their exam results. This was not going to happen until the following Sunday at church.

On Sunday, I learned that all twelve of us from the village passed the exam and we were going to the same school. It was time to celebrate. We had managed to stick together since middle school, and we had two more years at the new school. We embarked on preparations for the new school based on the instructions included in the letters our parents received. Because we were older, we did not need our parents to take us into town to shop. Supply-wise, we needed little compared to when we went to middle and high schools. Most of what we needed were personal items; the rest was to be provided by the school.

There were very few college-preparatory schools in the country at that time, one in the western part of the country called Tabora Secondary School, another in Central Tanzania known as Marianne Secondary School, and Korogwe, our new school on the eastern side, close to the Indian Ocean. These three schools were intended to produce female students who would go on to university. This was an important project that Tanzania undertook after independence under President Julius K. Nyerere, who was keen on addressing gender inequality in the workforce. The colonial government focused on men, relegating women to specific careers such as K-12 education, nursing, secretarial. Those who could not continue with higher studies returned to their respective villages in the hope of getting married and becoming homestay mothers and small-scale subsistence farmers.

Korogwe Secondary School was run by Episcopalian nuns, also from England. It was not hard to understand them because they shared the same culture as the Assumption nuns; they had similar religious philosophies and practices which, interestingly, did not persuade us to grow closer to them. Coming from Assumpta, we were already biased. We thought we were better because we were Catholics, what we were told was the only true religion. At that time, religious tolerance was not practiced. But we had to learn very quickly that we could not have a chip on our shoulders; we were not the majority.

The school was nondenominational, with many students of other faiths like Muslim and Lutheran. It was a government school, but the Episcopalian nuns, like the Assumption nuns, had an agreement with the government to run the high school and the additional classes for college-prep students. These nuns, like the Assumption nuns, started a middle school in Tanga, the capital city of the region. The high school and the preparatory school were about sixty miles to the west of the middle school and the nuns' main house or convent. The majority of the students at the high school came from their middle school. The nuns had an advantage too in securing spaces in the college-prep class for their own girls from the high school on campus.

We had been told a little bit about the school and the nuns before we arrived. Because we were the minority, we were a bit apprehensive. We strategized on how to adapt to the new environment and decided to arrive wearing our old school uniforms. We thought we would stand out from the get-go. When the train pulled up at the station, the welcoming party was surprised. They thought we were going to be in home clothes and not uniforms. But we had planned to show how proud we were of our high school and that we were unique. The welcoming party consisted of the principal, one teacher, and a student who was introduced to us as "big sister." In other words, she was the leader of all the girls and the link between the students and the administration.

The school was near the train station. We put our suitcases on our heads and started walking, following our hosts. When we arrived, the "big sister" led us to our dormitories. We were taken by surprise when we saw them. It was a step down from what we were used to. It looked more like middle school, but instead of single beds in a row, they were bunk beds. Additionally, one dormitory had students from all levels, ninth grade through the second year of college prep. Our identity was not the class we were in but the building we were in, named after important women like Anwarite, Madame Curie, Maria Goretti and so forth. Once assigned to a particular building, it remained the student's home for the duration of their time at this school. My dormitory was Madame Curie. Outside our dormitory was a framed picture of Curie with a biographic sketch of who she was. I was impressed and inspired since I was a science major.

At first, many of us were not happy at this new school. We were comparing everything with what we had before. We even wrote a group letter to our caretakers at Assumpta Secondary School, describing every aspect of the school and how we thought it was inferior to what we were used to. We did not like the uniforms, which were white blouses and orange skirts. We considered the material to be cheap and inferior in quality because it was not pure cotton but a cotton blend. We had hats that were part of the uniform but worn only when we went off campus, particularly to church. We could leave the campus on Saturday to go shopping and Sunday to go to church except for the students who were Episcopalian. They had their services on campus because they outnumbered the rest and because of the nuns. We were not allowed to wear headscarves, but Muslim girls were exempted from that rule when they went to the mosque. The wearing of scarves was the first incident at this school that landed the Catholic girls in trouble.

One Sunday morning, two months into the term, a group of us left campus to go to church about a mile away. The rule was that we had to be in uniform, and instead of a headscarf, we had to wear a hat from China made of synthetic fibers. They looked like the straw hats we saw in pictures of Chinese people working on farms. We all hated these hats and carried them in our hands. After a couple of trips to church, we decided to take scarves with us but conceal them until we were away from campus. For us, the scarf was a cultural symbol that identified us with the locals. Once clear of authority, we put on our scarves and walked on to church. Halfway there, one of the girls shouted, "Watch out, Sister Paula's car." Sister Paula was the principal. It was too late to take the scarves off and put on the hats, so we decided to scatter, running in different directions to hide. Four of us ran into a shop, with one trying to hide behind a refrigerator that stood by the door. I went behind the shelves and stood in the corner. The other two saw a door that opened to the back of the shop and continued through to hide outside the shop. Sister Paula parked her car and came inside the shop. Those hiding outside saw a chance to escape and ran to church. I could see Sister Paula enter. The shopkeeper was still in shock, wondering what was going on with the girls running through his shop, and now a nun coming in. Sister Paula asked him where the girls went, and he pointed to the back door. She did not go outside but decided to do detective work in the shop. She started looking around, and then peeped through the door that opened to the back. I thought she was going to go outside but she did not. She turned around and noticed the door of an old unused refrigerator that stood near the entrance was not completely closed. When she moved closer, she discovered one of the girls was trying to hide behind it. Her yellow skirt had given her away because it was visible through the cracked door. She could have fooled her if Sister Paula did not turn around to walk back. She told the girl to come out and she did. It was Martha from my class. Sister Paula looked at her and said, "Caught you beautifully, didn't I?" As if satisfied with her find, Sister Paula left through the front door without saying another word, but with Martha in tow. I was relieved but felt sorry for Martha. The shopkeeper came behind the shelf to let me know she was gone and that she took Martha with her. I came out and started running to church. I thought Sister Paula took Martha back to school, but she had taken her to church, dropped her off, and left to do errands. I arrived late at church and was surprised to see Martha there. After church, on our way back, Martha told us how Sister Paula discovered her and that she did not ask her about us. I told them I was hiding behind the shelves, and I saw how she searched the shop for us. I thought I was super lucky that she did not come behind the shelves. We did not know what to think, but waited for the worst to come when we were back at school.

  We lived in a cloud of fear the rest of that day and night. Because Sister Paula did not call for an assembly of the entire school that afternoon or come to the cafeteria during one of the meals, we began to think she had decided to let it go but would, at some point, make mention of the rule and give a warning about breaking it.

To our surprise, Sister Paula reserved her outrage for Monday morning assembly. The turn of events was more dramatic than we had envisioned. It was right after she finished announcements for the day that she called Martha to join her on the stage. The stage was a raised platform made of wood and stood imposingly five feet above the level of the floor facing the assembly. The stage was also used when the drama society performed a play such as the Shakespeare plays, like *King* Lear and *Julius Caesar,* or the Greek plays like *Oedipus Rex.*

Back to the Monday morning events. When Martha's name was called, we all sighed. She walked briskly to the foot of the stage and then climbed the stairs to join Sister Paula who paused for thirty seconds before speaking. She took off her reading glasses, wiped them, and then put them on again. She raised her head and started to speak.

"Listen to me and listen well. Look at this girl here. She has shamed our school, the Korogwe community, the West Lake Region, including her hometown of Bukoba, and the whole of Tanzania."

Those of us who were with Martha that Sunday morning were dumbfounded. We could not think of what she might have done after we scattered to hide from Sister Paula. We held our breaths, waiting for the unknown bombshell to be dropped, and wondering whether Martha was going to be expelled from school. Sister Paula continued, "Yesterday, I was driving to town, and I saw Martha and her friends walking to the Catholic church downtown. She had broken the school rule that requires students to wear hats with their uniforms but not head scarves. When they saw me, they all ran away, but Martha was not so lucky. She ran into someone's shop and tried to hide behind a refrigerator. When I caught her, I asked her why she was wearing a scarf instead of her hat. She said to me, 'but Sister, why?'"

We looked at one another still waiting for the bombshell. But there was none. We wondered what the shaming was all about. The nun continued, "Now, Martha and all the girls who were with her yesterday, they know who they are, must come to my office after this assembly to get their punishment."

I heard a student behind me mutter, "This is ridiculous." Sister Paula declared the assembly over. All the students left the assembly hall and headed to their classes. Those of us who knew we were guilty went to Sister Paula's office. The eleventh graders thought this was funny. They took Sister Paula's statement, "but Sister, why?" and turned it into a chant. They said it repeatedly, and their chant became louder and louder as they approached their classroom. It could be heard anywhere on campus and beyond. Sister Paula could hear it loud and clear from her office. She was not amused.

We accompanied Martha to Sister Paula's office. She looked angrier than she had on the assembly stage. We tried to show remorse, but we knew our punishment was already predetermined—working on the farm for some amount of time. Sister Paula looked at us, ordered us to change, and proclaimed three days of farmwork and no classes. We went to our dorms, changed into work clothes, and then went to the store to get what we needed to work on the farm. We went to the designated farm area and started working. It was mostly clearing the land and making raised beds for planting sweet potatoes. We were surprised to see another group joining us—the eleventh-grade students. They were going to be there for the day, a punishment for being rowdy and chanting. We thought this was unwarranted, but Sister Paula felt they were mocking her. Two days of punishment passed quickly. On Wednesday, we resumed our regular class schedules. We never broke the rule again, but we did not wear the hats either beyond the school gates. We wrapped them in plastic bags but put them on when we knew someone from the administration was watching.

The punishment did not make it easier for us to adapt to our new environment, and we did not make it a secret that we did not like this new school compared to Assumpta Secondary. We thought Assumpta was superior to Korogwe and, as seniors, we expected better accommodations—double-occupancy rooms. We resented having to mix with the high school students in the same dormitory, and we thought we were too old to sleep on bunk beds. Furthermore, we did not like the idea of going outside the dormitory to the bathrooms, even though they were right in an adjoining corridor outside the door that opened to the back of the building. There was a covered walkway with a wall around the back of the building for security. Along the same corridor were multiple large sinks that we used for laundry. Other things that were downers included the general landscaping. It had been in existence for about three years when we arrived at the school, which meant that it was started at the same time as Assumpta. We knew we were just being very fussy about everything.

The school had massive acreage to manage, which explained the overgrown compound. Because they did not have outside help for maintenance, the students had to assume that responsibility, cutting the grass with sickles. We were not happy about that either because we did not do that at Assumpta, where there were two lawn men with lawnmowers who cut the grass regularly. We only helped to keep the flower beds looking good. Our work was to study and study hard, not to maintain the school compound. We bragged about how civilized our high school was and that this was an inferior place. We became brats, complaining about anything and everything. We did not realize then, but the Assumption nuns had spoiled us rotten.

The food also was an issue for us. Granted, we had tea each morning, a novelty we did not have at Assumpta. We also had bread with margarine and marmalade, something we enjoyed at Assumpta on special days like Easter, and other religious feasts including the day the Assumption nuns celebrated the namesake of their order, the feast of the Assumption of Mary, the mother of Jesus. The feasting on that day was memorable because it included rare foods like bread or scones, rice with beef (occasionally chicken), and English pudding. We had to give the Anglican nuns credit for affording tea with milk, and bread with margarine every breakfast. But it was not enough to take away the fact that lunch and dinner were not impressive. It was corn or flour-based dishes with beans for the most part and occasionally beef or sardines. We grew sweet potatoes and cassava root (known in the Americas as yucca) that became alternate menus, especially on Saturdays and Sundays. On such days, a group of students was commissioned to peel the potatoes or cassava root on Saturday. Coming from the Kilimanjaro region, we believed that cassava root was poisonous and should not be consumed. Our culture ate yams, sweet potatoes or round potatoes, and bananas. The idea that we were going to eat cassava freaked us out, but those who came from the cassava-root culture convinced us that we would be fine because there were different species, and this one was not poisonous. The first time we ate it was for Saturday dinner. When Sunday morning came and we were still alive, we believed them. We were determined not to disclose this to our parents lest they freak out too and start looking for signs of health problems.

By the end of the term, we realized that nothing was going to change despite countless improvement suggestions we made to "big sister." For example, we suggested that those of us who were in the two upper classes should have more study time and fewer chores. We said the high school students, in grades nine through eleven, had plenty of time before their matriculation and we had done enough over the years to merit this entitlement. The school administration would not hear any of it. In fact, they thought the instigators of these requests were none other than the girls from Assumpta Secondary School, which had some truth to it. We were trying to create a new life that had some of the characteristics of our old school while the nuns wanted us to adapt to our new environment.

We settled down until we entered the last year of our studies, known as Form Six. This year, we were the elders. We had a lot of leadership responsibilities including as "big sister," school newspaper editor, chair of the debating society, and dormitory leader. Individuals in these positions constituted the student council. Each of them had a group of four to five students who were committee members who met regularly to decide on matters of student interest. Their discussions and decisions were presented to the council before the council presented them to the school administration.

My responsibility was the school paper, which did not have many issues that required administrative oversight. My committee members consisted of one student from each class. Our responsibility was to produce one newsletter each month featuring stories, poems, art, and funnies. I introduced a section that focused on jokes and made-up vocabulary items. Because some of the content poked fun at the nuns, the editorial board decided that it would be an insert included only in the copies distributed to the students and not the administration. We were not sure if they would share the students' humor. For example, the nuns were referred to as penguins because of what they wore. They had long white tunics with a black veil that was attached to a boxlike hat. The veil dropped down the back and was the same length as the tunic. This made the front all white and the back all black, just like penguins. One of the girls decided to write a story about a woman who lived in a community of silly people (CSP). She used this abbreviation throughout her story, and in between the lines, one could easily tell she was writing about one specific nun. Although the story was fictional and funny, the use of CSP was suspect because CSP was the title used at the end of each of the nun's names—an abbreviation for Community of the Sacred Passion. The editorial board agreed that we should keep the story but put it in the funnies section and as an insert, not in the main pages.

In our last year at school, we were no longer Assumpta students versus the others, but a unified group of seniors. I do not want to say we had influenced the other students to be independent thinkers, but each day we noticed a growing unity in what we liked and what we did not like. During student board meetings, we discussed changes we wanted to be implemented, and we made sure that the "big sister" delivered our message to the administration. Two issues arose in the first term. One was a move made by the administration to reduce the food budget. They increased bean dishes and removed meat, except on Sundays, and took marmalade off the breakfast menu. The board objected to this change, particularly because the school had bought excess bags of beans, which started going bad because of the humidity. Little bugs laid eggs in the bags containing the beans causing larvae and small black ants that appeared in the lunch and dinner dishes. The biology teacher told us that these bugs and larvae were harmless and, combined with the black ants, we were getting extra protein. This explanation did not fly, so the board called for a boycott of bean dishes and bread without marmalade. We were acting spoiled, but we were convinced we had the right to question the observation. Of course, there was some legitimacy in the issue with the beans, but the argument for marmalade was far-fetched. We knew not every family could afford bread for breakfast every morning, let alone margarine and marmalade. But we did not think about it that way. We were focused on "the right to question." The boycott plan was communicated to all the classes while the administration was kept in the dark.

On the day of the boycott, we went to breakfast as usual. The plan was to take the tea but not to touch the bread. The teacher on duty did not understand what was going on. One table started banging their plates chanting, "We want marmalade." The other tables picked it up and the whole cafeteria was chanting in unison, "We want marmalade." The teacher on duty called us to attention and ordered us to stand up for the end-of-meal prayer. We left the dining room without touching the bread. We were not sure whether the teacher on duty reported what we had done because the morning went on as if nothing had happened. At lunchtime, when we arrived at the cafeteria and saw that they were serving us beans, whispers went around that we were going to stage a walkout. Someone shouted, "Go." We all filed out of the cafeteria and went back to class without eating lunch.

The nuns and the administration did not take this kindly. The principal ordered the big school bell to be rung indicating that everyone must go to the assembly hall. When we arrived, the principal was already standing at the raised platform. She ordered us to sit down and started to lecture.

"I am not sure who you girls think you are. There is no doubt in my mind you just want to give me a headache. The school is trying hard to do the best for you. The least you could do is show some gratitude. We have telephoned the district education commissioner who will be here shortly. We are asking permission to send you home until you learn how to appreciate the government's good graces to you."

This was unexpected. We did not know what to make of it, but we had no choice. We had made our beds and now we had to lie in them. We sat quietly at our tables and waited. We heard the Land Rover as it entered the school gates. The principal went outside to meet the guest. After a short while, the teacher on duty came in and asked the student council to follow her. The district commissioner wanted to meet with them first. He was standing in front of his Land Rover and watched us as we approached him. He wanted to know why we had refused to eat the beans. He did not ask about the bread. Once he heard about the beans and the explanation from the biology teacher, he decided to go to the kitchen to see for himself. He saw the beans we had refused to eat first. Then he went to the store to inspect the dry beans. Then he came to the assembly hall to speak with us. He was an educator by profession and a parent too.

Based on how he reacted to the incident, we thought he felt our pain. He started by making general remarks about the importance of education and especially education for girls. He thanked the nuns for taking good care of us and for coming all the way from England to do that. He then addressed our grievances. He noted that he could understand why we questioned the changes in the budget and the fact that there was a problem with the beans that we refused to eat. He announced that the government was going to make changes and that the district would allocate more funds to the school to rectify the situation. He asked the student council to meet and make suggestions on a seven-day menu and that, effective immediately, marmalade would be back, and all the spoiled beans would be taken out of the store and sent to a pig farm nearby. He promised a truck load of new beans by the next day. We clapped and started chanting a praise song in Swahili about Tanzania.

Tanzania	Tanzania
Nakupenda kwa moyo wote	I love you with all my heart
Nchi yangu Tanzania	My country Tanzania
Jina lako ni tamu sana	Your name is overly sweet
Nilalapo nakuota wewe	When I sleep, I dream of you
Niamkapo ni heri mama wee	When I wake, I am at peace
Tanzania	Tanzania Tanzania
Nakupenda kwa moyo wote.	I love you with all my heart.
Tanzania Tanzania	Tanzania Tanzania
Ninapokwenda safarini	When I travel
Kutazama maajabu biashara nayo makazi	To see wonders business and places
Sitaweza kusahau mimi mambo mema ak wetu kabisa	I will never forget our good conduct ever
Tanzania Tanzania	Tanzania Tanzania
Nakupenda kwa moyo wote.	I love you with all my heart.

After the song, the commissioner clapped and shook hands with the principal and left the stage. He stopped to speak with the teachers while we filed out of the assembly hall to go to our classes. We were told lunch would be served at 3 p.m. We all wondered what it was going to be on such short notice.

The kitchen decided to make green vegetables to replace the beans. We felt we had won the fight, and we were glad that the commissioner tried to smooth things to accommodate the nuns' feelings. We were fortunate that the country had attained independence by then because the government's arm in the administration of the school made a difference. Otherwise, the boycott would have been viewed differently—as defiance by spoiled brats and outright disrespect toward the administration.

The student council met that evening and draw up a menu that we thought was balanced. We had meat on Sundays and Wednesdays; beans on Mondays, Thursdays and Saturdays; and vegetables on Tuesdays and Fridays. We had more days for beans because there were different ways of incorporating them in the dishes to create interesting varieties. We also requested meat on all religious and other major public holidays. We were not sure about the outcome, but the menu was well-received and implemented without modifications by the administration and the kitchen staff. Thereafter, the council and the administration had a cordial relationship, and many suggestions and requests were better received. At the student council meeting, we discussed the outcome of our attempted strike and agreed that the experience was worthwhile and that we had become better leaders despite the disruptions it caused. I was instructed to write a special editorial section in the next paper praising the administration and thanking the commissioner for supporting the school. We planned to send the commissioner a copy of that issue.

After this event, we gained an unlikely admirer. This was Miss Little, the mathematics teacher who thought that at our age we were courageous and bold in what we had done. Ms. Little was a young American teacher who had come to Tanzania as part of the Peace Corps program.

The Peace Corps was established in the United States in 1961,[1] six weeks after the inauguration of President John

---

1 cf. www.peacecorps.gov.

F. Kennedy. The idea was initiated by President Dwight Eisenhower, who had established an educational and cultural exchange program during his administration. Both programs took advantage of the decolonization of Africa and other regions. President Kennedy had an agreement with the government of Tanzania, right after independence, to have Peace Corps volunteers teach in secondary schools across the country. Korogwe High School was one of the few schools that had an opportunity to have a Peace Corps teacher assigned.

Ms. Little was the only math teacher for the entire school, a general trend, except for English and religious studies. This kept the teaching staff minimal but was a lot of work for one teacher considering we had six classes with thirty-five students in each. While English had two periods a day, the other subjects were taught twice a week. We had one English teacher per class to teach grammar, reading, composition, and literature. Swahili and religious studies were taught once a week, on Fridays. Swahili was taught in the morning, freeing the afternoon for religious classes because the Muslim girls had to go to the mosque outside campus. The girls in the college preparatory classes did not have to take many classes. We had a slate of courses we could choose from depending on whether one was interested in hard science (math, physics, chemistry, and biology), social science (biology, math, geography), or general studies (geography, history, and Bible studies). All students had to take civics and language.

Miss Little came to campus the same year as we did but a month later. It was interesting to see what she had brought with her to help teach. Her assumptions about Africa were like those depicted in the movie *Tarzan,* a jungle with no houses and no classrooms. She brought tons of visual aids, including portable writing boards, pens and pencils, bundles of writing paper, and more. She believed that her classes should be outdoors under a tree. She brought a tent that she would use in case it rained, a folding table, and a chair for herself. She was surprised to find a well-built school and kids dressed in cute uniforms. She also was surprised by her modern house that had two bedrooms with self-contained bathrooms, a living room, formal dining room, and a fully furnished kitchen. She also had an assigned housekeeper to help her with cleaning, laundry, and keeping her garden. If she preferred, she could make simple meals like breakfast and afternoon tea. The housekeeper, who was paid by the school, did not live in the house but came in the morning around seven and left at five.

We loved Ms. Little. She was funny, and liked to sing, and hang out with us. She was the only teacher at school who wore high heels every day. We wondered how she could walk so fast in them on rough, unpaved roads. Rain or shine, Miss Little was always in high heels. We thought her walking speed matched her speaking speed. She spoke so fast with her American accent that it was hard to understand her at the beginning, as it was for her to understand us with our acquired British accent. As time passed, we got used to the accents and spelling differences. The school paper nicknamed her Ms. Jet, liking her to the jet planes we saw flying over our school from time to time. She was friendly and liked to hang around the senior girls at school more than with the other staff members. We never asked her how old she was, but we guessed she must have been only a few years older than us since she had just graduated from college. She told us a lot about America and was willing to answer our questions no matter how silly they were. She taught us American songs. I remember the song and poem telling the story of Hiawatha. I also remember "Michael Row the Boat Ashore" that we liked to sing when working in the garden because it had a good rhythm as the group raised their hoes in the air and brought them down in unison to turn the soil ready to plant something new.

"The Song of Hiawatha"	"Michael (Row the Boat Ashore)"
By the shores of Gitche Gumee,	*Michael row the boat ashore, hallelujah*
By the shining Big-Sea-Water,	*Michael row the boat ashore, hallelujah*
Stood the wigwam of Nokomis,	*Sister help to trim the sails, hallelujah*
Daughter of the Moon, Nokomis.	*Sister help to trim the sails, hallelujah.*
Dark behind it rose the forest,	*Jordan's river is deep and wide, hallelujah*
Rose the black and gloomy pine-trees,	*And I've got a home on the other side, hallelujah.*
Rose the firs with cones upon them.	*Michael row the boat ashore, hallelujah*
Bright before it beat the water,	*Michael row the boat ashore, hallelujah*
Beat the clear and sunny water,	*Michael's boat is a music boat, hallelujah*
Beat the shining Big-Sea-Water.	*Michael's boat is a music boat, hallelujah.*
	Michael row the boat ashore, hallelujah
	Michael row the boat ashore, hallelujah
	The trumpets sound the Jubilee, hallelujah
	The trumpets sound for you and me, hallelujah.
	Michael row the boat ashore, hallelujah
	Michael row the boat ashore, hallelujah
	Michael row the boat ashore, hallelujah
	Michael row the boat ashore, hallelujah

We were fascinated by her culture especially her love of dogs—not just any dog, but her cute puppy that she brought with her. It was a little poodle, white with black spots. She told us it was her baby and she had named her Violet. It was a nice name, but the relationship, interpreted in our culture, was strange. One of the girls asked Miss Little: "How can a dog be a human person's child?" From the way Miss Little looked at the girl, we knew that Violet was special to her, and we needed to understand and accept that reality. Some of us in math class took turns babysitting Violet and learned her routines, such as brushing her fur, tying ribbons on her ears, and putting cute booties on her tiny paws. Although we thought this was strange, we respected Miss Little because Violet was special to her. The biggest surprise was when we were invited to Violet's birthday. The whole math class was invited, and the event was held at Miss Little's house. She had decorated the living room with balloons and there was a cake on the dining table. After singing the happy birthday song to Violet, Miss Little helped Violet blow out the candles and cut the cake. After feeding her, she gave each of us a piece of cake and a cup filled with Kool-Aid, which Miss Little made. Violet had a taste of the drink from Ms. Little's cup. We tried not to show indifference because we did not want to disrespect our host who was also our beloved teacher.

We thought this was a strange cultural relationship with a dog, but we did not want to judge Miss Little. What we wanted to take away with us was that her culture was different but that, as a person, she had a big heart. We were grateful that she made sacrifices to leave her family and come across the globe to Africa to teach us as a volunteer. What her mission entailed was not something we wondered about. She was our teacher and had one child, Violet the dog. Some of us were curious enough to ask whether we could visit her home country to see what it looked like. Little did I know that I would have that opportunity eleven years later.

I credit Miss Little for passing me on my advanced-level math examination. Like in high school, the advanced-level exam, popularly known as A-Level, was shipped from England to all the colonies. We were the last group to take the A-level exam, after which the government replaced it with its own national O-level and A-level exams. Both exams were traumatic to students, not because they were coming to us by boat, but because they were alienating. We had no idea what England looked like, let alone the people who were setting these exams without knowing the local conditions and assuming the test pertained to all students regardless of where they lived and the resources available to them. In addition, these exams were in a language that was not even second but third to many of us, considering that we spoke a home language determined by our ethnic group, a national language—Swahili—and then English, the language of the colonizer.

Ms. Little was committed to having as many of us as possible pass the exam. She agreed that it was weird that a foreign country was administering an exam to test students who they did not know or teach. Because there were only ten of us in the class, Ms. Little spent countless hours with us on review exercises before the exam. The extra hours paid off. When the results were announced, all of us had passed at an acceptable level. I needed math and physical science studies because I wanted to study architecture at the college level.

After the terminal exams in November, we headed home to wait for the results in late December. The only difference from high school was that there was a preselection process for college entrance through a national mock exam in the middle of the year. The preselection took away the cloud of uncertainty at the end of the year after the A-level exams. From the results of the mock exam, I was preselected to join the university after six months of national service. Waiting for the final exam results during the holidays was less traumatic compared to the time after high school. This was huge because we had one university in the entire country and getting in was a great honor. Ten of us had been preselected. It was hard to say good-bye to each other at the end of the term, knowing that we would not see some of our classmates the following year, and we did not know what would happen to them if they did not pass the A-level exam.

National service was a new requirement by the government for all students who had received four or more years of government-subsidized tuition. After independence, the government eliminated tuition for high school through the end of college. Based on this, I was obligated to show my gratitude for free education by participating in six months of national service before entering university or public service.

The Christmas holiday went quickly, and in January we started to prepare to go to our assigned stations for national service. My station was on the northwestern side of Tanzania on the shores of Lake Victoria. As I was preparing to leave, a tragedy befell my family. My mother was involved in an accident. It was a Sunday morning and the first service at our church had just ended. The service, which started at six in the morning and ended at eight, was reserved exclusively for women and was followed by the children's service at nine. The third service started at ten and finished at noon and was exclusively for men (at least eighteen and over).

The story was that a motorcyclist was riding down the main road, which was crowded by people coming from the service. The church was on the main road, which connected many of the villages from where the faithful came. Sharing the road with buses and other motorized vehicles was tricky because of the number of people streaming from the church as they walked back home. Drivers honked to warn the pedestrians to yield. If the pedestrians didn't heed the warning, an accident was likely. We do not know if the motorist warned the pedestrians or if he just plowed into the crowd. There was an incline, and it is possible the motorist failed to control his brakes because a witness said he hit my mother who had already moved to the side of the road. My mother hit her head on the gravel road and suffered a severe concussion.

When this accident happened, my sister and I were already inside the church and my brother Ladi, who had just been ordained a priest, was giving the homily. The news about mother's accident was relayed to him after he finished delivering the homily and he left right away to check on her.

After the accident, my mother was taken to a clinic near the church that was run by Catholic nuns. Because she was unconscious and the clinic was unequipped to help her, they decide to transfer her to the district hospital in town after Ladi arrived at the clinic. I am not sure how transportation was arranged; I suspect the church provided it and Ladi was able to accompany her. My father was informed of the accident before he left home for the third service at ten and left immediately to assess the situation.

My sister and I had no clue what had happened until after the service. News about my mother's accident was traveling throughout the village. As we were walking home, a woman approached us and said she was sorry about what had happened to Mother. We were shocked and asked what she was talking about. She realized we did not know and started to explain how our mother had been hit by a motorist. We sprinted home, wailing all the way. We were in shock and lost about what to think or do. We did not have the whole story; we did not know how bad the accident was or whether Mother was dead or alive. When we got home, we must have continued crying uncontrollably for an hour or so until our neighbor came to the house to try and comfort us. She was there when mother was hit and assured us that she was going to be fine after the doctors had a chance to check her out. We settled down and started making lunch with heavy hearts. Later that afternoon, Ladi came back from town and gave us an update. He knew mother was in critical condition but tried to assure us that she was going to be fine. I could tell there was something weighing on his mind, but he was not going to tell us much more. Father did not return home until later that evening. I did not expect him to reveal much to us either. He brought more food for the family and gave my sister and me instructions on how to manage things until Mother returned. He indicated that it was going to be awhile.

Mother left us with a six-month-old baby to take care of. It dawned on me that I was responsible for this infant and from now on I was the surrogate mother. I knew how to prepare the formula for the baby and most of the baby food my mother made. Everything was natural and used local ingredients like milk, maize flour, potatoes, green vegetables, and green bananas. It also dawned on me that it was not going to be possible for me to leave to go to national service before my mother's condition improved. I started finding out what my options were. I spoke with Ladi about it, and he suggested asking the head priest at the parish to offer some suggestions. He took charge of that and in a couple of days brought the news that the government had allowed me to postpone my departure for a month.

My father tried to keep things as normal as possible at the house, but my mother's absence was inescapable, even the cows and the goat at the house sensed her absence. My father visited my mother in the hospital every day and mobilized a support system from our aunts. They took turns staying at the hospital with Mother and came to the house to make sure we were doing well. They provided support feeding the animals, gathering enough grass to last at least a few days to alleviate the pressure on us. When my aunts could not be at the hospital with my mother, my father took me to spend the night with her. That is when I realized the gravity of the situation. My mother was still unconscious, mainly because the doctors gave her Valium that kept her sleepy most of the time. I watched her sleep all day. At night, the Valium was less effective, and she would lay awake confused about where she was. One night, I fell asleep and was not aware that she had woken up and wandered out of the hospital room. She went outside and started wandering around. I woke up suddenly and realized she was gone. I ran outside, terrified, looking everywhere for her. Then, I saw a shadow near the mortuary building. I ran to it and, sure enough, it was my mother. She was talking to herself, unaware of where she was. I grabbed her by the hand and started walking her back to her room, trying to talk to her at the same time. As we entered her room, a nurse came up behind us and asked what was going on. I told her what had happened, and her response was to give my mother more Valium, which sent her into a deep sleep.

The next day, my eldest brother Wenceslaus came from Dar es Salaam where he was working for the post office. I told him what had happened. I could see it disturbed him. He told me he was going to town and would be back shortly. I do not know where he went or what he did, but I suspected he must have talked to some of his friends about the situation. When he returned, a friend accompanied him. They planned to transfer Mother to a private hospital that was about twelve miles out of town in a subdivision called Machame. The hospital, Nkwarungo, was run by the Lutheran Church. He told me to go home, indicating that he was going to be with Mother for a while. I left and went home without knowing his plan. My father was shocked to see me; he must have thought the worst had happened. I told him that Wenceslaus had arrived from Dar es Salaam and was taking care of Mother, which calmed him down.

With the help of his friends, Wenceslaus successfully transferred Mother from the government hospital to the private one. Rumor has it they did not seek the doctor's approval but abducted my mother from the government hospital and checked her in at the private facility. We were told later that as soon as she arrived at Mkwarungo, she underwent a battery of tests and the doctors discovered she was dehydrated, and her tongue had turned green from not eating or drinking for an extended period at the government hospital. The doctors switched into high gear to save my mother's life. After a few injections, she regained consciousness that evening. She started talking and asked about her baby. They kept her on intravenous liquids and by the end of the second day she ate her first meal—soup—raising the doctor's hopes. Wenceslaus came home on the third day and organized for the baby, Adeline, our last-born brother, to join his mother in the hospital. According to the doctors, this was psychologically beneficial and a major contributor to my mother's rapid recovery. I prepared the baby to be reunited with his mother, but I did not accompany him on this journey. My father and Wenceslaus took him. I was relieved that Mother was getting better after a month in a coma because I was concerned that my excused leave from national service would end, and I would have to go before I knew her prognosis. We received reports that she was making good progress and would be released soon to go home. Unfortunately, it did not happen soon enough, and I had to leave for my national service post.

I left feeling optimistic because the baby was with his mother and my mother was aware of her surroundings. Because my sister was still at home waiting to go back to school after the Christmas holiday, I felt the homefront would be manageable. She just needed to prepare simple meals and do general day-to-day chores because my father had to spend most of the day at school teaching. But a home of just men was going to be interesting when my sister finally left for school. Because Wenceslaus had taken a short leave of absence from work and was going to be in town for a little longer, I was confident that things would be under control. Most of all, I was confident that the worst was over. We had been spared the trauma of losing Mother.

The trip to my national service post was uneventful. It was long but interesting because it was my first long trip to the region around Lake Victoria to a place called Sengerema. Compared to my home region, this area is drier, warmer, and heavily affected by strong winds because the land was mostly flat. The lake has strong currents because of its proximity to the Rift Valley that runs from Ethiopia to Tanzania, creating occasional tropical cyclones. During our stay in Sengerema, we experienced one of these tropical cyclones. It occurred in the middle of the day without much warning. The camp leader rang the warning bell while the section leaders hurried us back to our tents to seek shelter. Several people were slow in getting to their tents, and we looked in awe as they struggled to get to shelter while the wind kept pushing them away. The iron-sheet roof over the storage shed was impacted by the strong wind, intensifying our fears of danger in case the pegs holding the tents gave way to the severe weather. Although this lasted a mere fifteen minutes, it seemed like an eternity. When it was over, a blistering sun came out and in half an hour the place was dry and dusty. If we did not have objects, tree branches, and other debris scattered all over, no one would know that a cyclone had just passed through.

I had only two months at the station because I was a month late. We were required to be at one camp for three months and another for the final three, a total of six months of service. I would have felt like an alien when I arrived at this camp if I did not know anyone. Luckily, there were a few girls already at the post from my school. They were happy to see me and helped me adapt to the new environment. They had valuable tips on how to survive gender put-downs by the men in charge of the post. There were fewer women than men, which was inevitable because there were fewer women in high school and college preparatory schools. Activities at the post included farming, keeping chickens, and managing the environment. We lived in permanent camps with no running water. Most of the water we used came from a single pipeline that drew from the lake and harvested rainwater that was used for watering and laundry. Because the pipe had limited outlets, we had buckets to collect water as needed for various activities.

We were very busy between breakfast and lunch, but did not do much in the afternoon because it was too hot. We used that time for personal maintenance, sports, or to tour nearby villages. This post was in an area called Butiama, the village where the president of Tanzania was born and raised. One day, out of curiosity, we decided to walk to the location. We heard that his mother and one of his brothers lived there and that the president had a small house there that he used whenever he visited his mother. Several of us decided to take this adventurous tour but turned out to be unwelcome. The sheer size of our group took the residents by surprise as we approached the house. At that time, there was no need to request a visit in advance. One could show up at many official or historic places without an appointment or invitation. Strangers were given a courteous reception and the benefit of the doubt until proven otherwise. We made the trip thinking we could take advantage of this and didn't hesitate to invite ourselves to the president's village, let alone his home. When we entered the compound, the president's brother was sitting outside the house carving a cane for his mother. He saw us approach and without hesitation called out several times, "Welcome, welcome, welcome, you are all welcome."

We politely accepted and moved closer to the house to where he was seated. He asked what he could do for us, and one of the girls responded by introducing ourselves and telling him that we were with the National Service Corps at the post nearby, and we just wanted to pay our respects to the president by visiting his village and home. He seemed happy and proceeded to tell us about the home. This was where the president was born and because he was a son of a chief of a small ethnic group called the Zanaki, they had a lot of property that was distributed to the chief's children and grandchildren. He said that the buildings in the compound were not original but were built by the president when he was still a teacher. By all standards, these buildings were average—small but with modern features like brick, iron-sheet roofing, cement floors, and good paint. Our host stood up like he was going to show us around when an elderly woman emerged from the house. We assumed this was the mother. She asked him in Kizanaki, her native language, who we were and what we wanted. Because Kizanaki is a Bantu language, it was easy for us to pick out some of what he said in response. The mother had a terrifying look on her face, a warning that she was not happy with us inviting ourselves to her compound. She continued to talk to her son, instructing him on what to tell us. Her message was that Butiama was a big village, and we could go anywhere we wanted to visit but not this compound. Once her son relayed that to us, we had no choice but to turn around and head out of the compound as fast as we could. We wished the president were at home because he would have been happy to see us considering that we were part of the National Service Corps in his home village. He was a very warm person, a good teacher, and was referred to as the father of the nation. Back at the post, we shared our adventure with those who had not gone. Our experience served as a warning to them not to venture there.

Two months passed quickly. The experience was good except for the mandatory early-morning exercises that included a two-mile run and a one-time cross-country marathon that the camp organized to start at two in the morning. By the end of it, we all had blisters on our feet and our entire bodies were sore. The nice thing was that the next day all major activities were canceled, and only small chores assigned, leaving us with a lot of time to rest.

The three months at this camp prepared us for the next post, which was in a region on the eastern side of the country. This post was closer to home, and the weather was cooler without much risk of cyclones. We were bused to the new location, an entire day's road trip. The facilities were similar except that the post administrators were tougher on us than the first group had been. Many of them were senior national service staff and ran the post like a boot camp. We had fewer hours of downtime because there was a lot of work including construction, livestock, farming (corn, beans, and millet), digging trenches for irrigation pipes, and preparing animal feed. The day started much earlier than at the first post, six in the morning as opposed to eight o'clock. Because the climate was cooler, there were activities scheduled most afternoons except Saturdays and Sundays when we could go to town and hang out. The post authorities were more professional than at the previous camp and treated women with dignity. Because we were busy all the time, the twelve weeks went by very quickly. At the end of the program, the camp organized a celebration that included a visit by the minister of youth and culture who gave a graduation speech. We were given a certificate of completion and our final pay. Each month, for the six months of national service, the government gave each of us sixty Tanzania shillings (about $9). It was a lot of money at that time for a youth who had never received a paycheck.

After the ceremony, we started packing up to go home with one goal—to prepare for college. Several of us from the same national service post were heading to the University of Dar es Salaam's main campus in the capital city of Tanzania. This was the only university in the country, but it had an off-campus branch that focused on agriculture in a city that was about 108 miles away from the capital. Three girls from our school had been selected to study agriculture at this campus, and one student had been selected to go to Uganda to study at Makerere University.

My family was happy to see me after six months. I was happy to see my mother who had been hospitalized for three months due to the motorcycle accident. I had left while she was in the hospital. She had healed completely without any aftereffects or other health issues. I did not have much time at home. I was expected on campus by July 11. I had saved my national service stipend and could buy nice clothes and shoes for college. We did not have to wear uniforms anymore but had been warned that the attire was business casual. This meant that every day we would be dressed for class like we were going to church. At that time, girls did not wear trousers, only skirts and dresses that could be short, midi, or maxi. I had to find a good tailor who knew different styles that would look good on me. I made sure I had enough outfits that I could mix and match to cover a week's worth of classes. I knew I could buy more once I saw what others had.

— 6 —

College Life

When we joined the University of Dar es Salaam, it was emerging from a difficult time. There had been strikes that forced the government to close the university and send all the students home. The strike was precipitated by the government's institutionalization of national service. The students did not think they should have to do national service just because they received free tuition. They claimed it was the responsibility of the nation to educate its citizenry and that they were destined to expand the public/government service upon graduation, and that was enough payback. It was this resistance that made the government change its timetable for national service by requiring it upfront before entering college. My group was the first under the revised policy requiring national service as a prerequisite to college entrance. The policy lasted for several years and then it was modified for medical school students because they were required to do seven years of study to become doctors while the other degrees required only three.

The university was located on the west side of downtown Dar es Salaam in a subdivision called Ubungo. The public nicknamed the place "The Hill" because it sat majestically on a hilltop that was visible from the major highway that connected Dar es Salaam to other cities. The buildings were mostly high rises and also visible from the highway referred to as Morogoro Road after one of the major cities the highway passes through, and which served as an interchange for all transportation heading to the western or southern parts of the country. Morogoro Road was, at that time, the only main surface exit out of the capital city of Tanzania. We used to joke that if there were a disaster in Dar, we would all die because the road could not withstand the stampede. To some extent, it is still true today except that a new major road was constructed passing through Bagamoyo, the historical coastal town and the first capital during German rule. But it is not the shortest route when trying to link to the southern and western states. The new Bagamoyo Highway goes through the northeastern regions before it curves to the central part of Tanzania through the new capital city, Dodoma, another midpoint for all junctions leading to the southern, western, and northwestern regions. The city of Morogoro was and remains a strategic place for the University of Dar es Salaam because at that time it hosted the Extension campus that was reserved exclusively for students studying agriculture and veterinary science. Eventually, the Extension campus became a full-fledged University of Agriculture, giving Tanzania its second institution of higher education. To date, Tanzania has several government and private universities that increase educational opportunities for its citizens.

When we arrived on campus, we were impressed. The buildings were modern and attractive. They were set up to meet international standards with two students to a room. The men's halls, designated Hall One and Hall Two, had up to five floors each, and there was an elevator. The electricity was reliable with no frequent losses of power that could have resulted in people being stuck in the elevator.

Each floor in the men's high-rise buildings had ten rooms, two students in each, and a shared bathroom at the end of the hall. Each bathroom had at least five shower stalls and three toilets. The female students had one residence hall, named Hall Three, set up differently compared with the male students' building. It was not a high-rise but an L-shaped structure with only two floors. Each floor had double occupancy rooms and a well-equipped kitchen where one could make a simple meal. At the end of the hallway, there were shower and toilet stalls. Both the men's and women's rooms were well equipped with a bed, mattress, pillow, and closet that was big enough to accommodate the clothes of two people. There was also a bedside table with drawers for small items. We did not have to bring our own bedsheets. The government contracted with a company that supplied each student with linens. The company picked up the dirty linens and dropped off fresh ones every Wednesday. All we needed to do was strip the sheets and leave the dirty laundry on our beds. The company brought the fresh linens and made the beds—hotel-style service.

The women's hall had added security. There was a matron assigned to the hall with responsibilities that included providing security and counseling to the girls who lived there. There was a feeling that the girls were vulnerable and needed protection. There also was a matron and guard who were stationed at the entrance to the hall from six in the evening to six in the morning. The matron and her family occupied a beautiful single-family house near the only entrance to Hall Three. Although there were no strict curfews, we were expected to be inside the premises no later than 11 p.m. If anyone came home later than that, they would need to get the security guard to let them in. The matron would be informed and might have a discussion with the individual the following day.

For those of us who came from sheltered environments, the setup with a hall supervisor was welcome. It was the first time since age eleven that we were in a coed educational system. We had a rough experience in national service where we had to endure gender-related put-downs. We prided ourselves on our endurance, a sign of the strong leadership and good upbringing by the female teachers and nuns in middle and high school. But we also realized that being sheltered had made us vulnerable, and we had to learn how to navigate in this new environment.

The challenge was to be independent and have good judgment even at the risk of being rejected. We retained sisterhood, looking out for one another, at least during our first few weeks on campus. We walked in groups when going to the dining room, classes, social hour, and off campus. We looked up to the older girls, who were in their second and third years, who came from the same prep school. We watched how they carried themselves, dressed, and behaved on campus. We adopted their good habits and tossed out the bad ones. They told us about obnoxious boys on campus, specifically one who was a good journalist and artist. He was nicknamed Mr. Punch. He wrote long essays about anything and everything that happened on campus and posted them on tree trunks. Sometimes he created obnoxious misinformation and stories about girls he did not like. Now we would label it as "fake news." We knew the student government could not stop him, but we were not sure whether the school administration knew about it and, if they did, why they did not do something.

One essay caught the attention of all the girls in Hall Three. By breakfast the next day, the whole campus knew about it because Mr. Punch had posted it all over campus with vivid drawings of how the girls had reacted to the incident.

It happened around eleven at night when most girls were already in their rooms preparing to sleep. One girl was coming from the bathroom and spotted a naked man walking along the corridor. She ran back to the bathroom and locked herself in one of the stalls. After a few minutes, she came out again and he was gone. She told her roommate and they knocked on the other doors to inform the girls about this incident. As they were planning to go tell the matron, they heard a scream from the top floor. They picked up whatever they could lay their hands on, including brooms and kitchen utensils. They rushed upstairs and came face to face with the naked man. They started chasing him, screaming at the top of their lungs. The guard and matron came rushing in to see what was going on. The guard grabbed the young man and wrapped his long coat around him to cover him up. He took the man away while the girls were still shouting at him with their brooms held up in the air.

What surprised us was the story that Mr. Punch distributed around campus. We wondered how he knew about it and how he got the drawings that depicted the scene with the girls chasing the perpetrator with brooms and pots and pans. The girls were in their nightgowns or robes, another detail Mr. Punch could not have known unless he was at the scene. We started thinking that the naked man must have been Mr. Punch. Since we could not identify him in the dark, it was impossible to pick him out if we saw him on campus. It was determined that he had come to this wing of the building through a back door that the girls had neglected to lock. The night guard and matron set a schedule where the back doors were locked before nine and the front door at eleven. There was also a rule that prohibited anyone from opening the back door for anyone coming in after-hours.

Campus life was full of unprecedented privileges. In addition to the linen provision, the dining room looked like a five-star restaurant. Each table was covered with well-pressed white linen tablecloths. The tables were arranged to seat ten people with spoons, knives, forks, and linen napkins. The food was served buffet style, but there were waiters who dished up the selected choices onto a porcelain plate and placed it on a tray at the end of the line. There, a waiter picked up the tray and escorted you to your table.

Breakfast included bread, eggs, and sausage while lunch was an assortment of dishes including vegetables and dessert. Dinner was like lunch but included soup. The waiters brought out the desserts and picked up the dishes. In addition to these meals, there was a four o'clock high tea, which was served with a variety of snacks.

Experiencing these luxuries was an awakening to us newcomers, and we could understand why the government was harsh on our predecessors who went on strike because they did not like the mandated national service. The government announced on national radio that the students were being sent home to experience the differences between living at their home and living on the Hill with free tuition, accommodations in high-rise buildings, indoor plumbing, electricity, free food, an annual stipend, an allowance for books and travel, a well-maintained campus with manicured lawns, first-class food, laundry services, and other amenities that were unique to the privileged university students. The government claimed that this was taxpayers' money and the least the students could do was appreciate the sacrifices the country was making to give them a first-class education.

The campus had a student recreational center with areas where students could relax after classes. It was managed by the student union, and a small bar sold alcoholic and other nonalcoholic drinks. The bar was always full at five in the afternoon until closing at 9 p.m. There also was a grocery store that sold a lot of the students' preferred snacks, drinks, and toiletries. In addition, there was a bookstore where we got all our textbooks. The government gave us an allowance for these books that was automatically paid to the bookstore management. After a student had found the books needed, they showed the cashier an identification card issued by the university to check out at the register. Next door to the bookstore was a bank and a post office. All these facilities served the needs of the students making it unnecessary to make trips to the city.

For recreation, there were intramural fields for soccer, volleyball, and track and field, along with a swimming pool. There were in-house competitions from time to time, but mostly the Hill was devoted to classes that started at eight in the morning and ended at four in the afternoon Monday through Friday except Wednesday when classes ended at noon. That afternoon was reserved for personal maintenance, short trips to the neighborhoods or the capital city. In the evening, the student union organized a social event, mostly showing films. Occasionally, Wednesday afternoons were used for public debates on hot political topics. The speakers were professors from the political science department or invited guests. The debates following a public lecture or roundtable discussion were the highlight. At that time, popular topics were democracy, socialism, capitalism, and the Cold War. No one wanted to miss these events because they were a major source of information about the world and a learning opportunity across all disciplines.

The weekends were interesting too. The university provided facilities for different religious groups, eliminating the need to go off campus for services. On Saturdays, some students walked around the neighboring communities and markets. Most students stayed on campus to study, do homework, or personal maintenance. On some Saturdays, the student union social affairs department hired a band to play and the students danced till midnight. In my second and third years on campus, I was a member of the student council and was appointed as the minister for social affairs overseeing all the social activities on campus. It was a challenge when I had to order a film because I did not know much about movies that would be of interest to the students. I relied on the title and the advertising picture. I was lucky because I never went wrong with my selections; the students loved each movie I picked. My biggest hit was when I chose a Clint Eastwood movie, *The Good, the Bad, and the Ugly*. We showed it twice due to popular demand. There was a small fee at the door for both the dance and movie nights. The revenue was used to offset some of the costs of the social events while the bulk of it was paid for by the student council.

As the minister for social affairs, I also was responsible for outreach. I had the opportunity to pick the projects my committee could work on. We decided to partner with the

University Primary School near campus. Many of the university employees had children who attended this school. In addition, the school had a close relationship with the Faculty[2] of Education; it was one of the schools where aspiring teachers went for practice. I had been at the school for my practicum because I was a double major student in Linguistics and Education.

The goal was to do a project that addressed a specific need at the school. I visited with the principal to assess the school's needs. The principal had a long wish list, but I chose one, a library project. I intended to raise funds to create a small library and stock it with children's books. I did not know where the funds would come from, but I was determined to find the resources for the library. The social affairs board thought I was crazy, but it did not deter me from pursuing my plans.

I drafted a plan describing the rationale for a library to benefit kids. It did not matter how much money I raised because I told myself the mission was to secure space for the library at the school and stock it with books from the funds collected. This would be the foundation on which the school could later build a more comprehensive library by appealing to the Ministry of Education and the community. I was convinced that the project was noble, doable, and attractive enough to move people to contribute.

With my plan in hand, I wrote a cover letter, made several copies, and started a fundraising crusade. I visited the offices of the university's administrators and requested a meeting. At each of these meetings, I introduced the proj-

---

2 Faculty is used instead of college. Thus, Faculty of Education is the same as College of Education.

ect and shared the plan and the cover letter that requested support from the individual officer. I was surprised by the overwhelming support I received. I also discussed the plan at the student council meeting and requested that they devote 10% of the revenue generated by social events to the project. After extensive discussions, the council approved my plan and pledged the suggested 10%. I decided to visit with the district education officer assigned to that school and shared my plans with him. I was surprised, too, that he warmed up to the plan and personally contributed to the project. He also promised to speak to the principal of the school to ask him to send letters to each parent with a child at the school with a suggested amount to contribute. With this support behind me, the project became much easier than I thought.

Three months later, I had collected enough money to start the project. The university assigned one of its accountants to work with me and to open a special account for the money collected. I also was assigned a librarian from the university's main library to work with me to set up the facility. I was anxious to complete the project that year, my junior year, because I did not want it to interfere with my senior obligations at the university. In the end, we were able to secure books—fiction and nonfiction—for the library. My love of books was an asset because I had an idea of what children would find interesting. The district commissioner came out for the library's grand opening and announced a special fund from his office to pay for a part-time librarian. That was my happiest day; mission accomplished.

My three years at the university went by quickly. On arrival, I was interested in studying architecture. The subjects I had studied and tested for at A-level qualified me to major in architectural studies. However, that was not the government's intent that year. There was a major push to train teachers to meet the campaign for universal education. The president had promised 100% literacy throughout the country in ten years. It was announced at the university that the options for that year were law and education. The university had offered available degrees to select from—science, law, and humanities. With architecture off the list, I was left with two choices, either a bachelor of science degree with a minor in education or a bachelor of arts with a minor in education. I knew I could keep my love for math in the science degree, but I would also have to take physics and chemistry. I liked physics but not chemistry. I tried to substitute biology for chemistry, but that combination was not available. I had no choice but to select arts degree. In the bachelor of arts category, there were courses to choose from like geography, history, linguistics, English literature, and Swahili literature. The minor in education had a set of required courses that included psychology, principles of education, primary education, secondary education, and administration. There was one required course, Development Studies, which was primarily political science, socioeconomics, and rural and urban development. I decided to take the BA with the education combination. I chose Swahili literature because emphasis on Swahili was minimal in colonial education and the president had just declared Swahili as the national language of Tanzania. And I decided on primary education because that program was for those who would teach at a primary school. This program allowed me to work as a teacher trainer. I did not want to teach at a high school. At that time, the university had a few foreign professors, mostly from England or other Commonwealth countries, and one from Jamaica. There were several African professors from Tanzania, Kenya, Uganda, and South Africa who were educated in England, Australia, and the former East African University. The East African University was established by Kenya, Uganda, and Tanzania but was later dismantled after independence to create independent institutions for each country (the University of Dar es Salaam in Tanzania, University of Nairobi in Kenya, and Makerere University in Uganda). There was one professor from the United States, an American married to a Tanzanian. She taught education psychology.

I chose to study linguistics because it was the first time I had heard about that subject. I thought it must be sophisticated and challenging like physics or mathematics. The first day of linguistics class was interesting. The professor was Mr. Hill, an Englishman. The focus was on English linguistics, a mixture of phonetics (IPA—the International Phonetic Sounds), phonology, morphology, grammar, and communication skills. I was surprised when I went to England years later to realize we had only touched the surface in my undergraduate linguistics class. There were other aspects that we did not cover such as syntax, sociolinguistics, psycholinguistics, artificial intelligence (computer linguistics), discourse analysis, and historical linguistics. In England, the emphasis was still on Indo-European languages, and any analyses and theories were based on this with little or no bearing on the 99% of non-Indo-European languages. It was not until then that I started to question the notion of universal language theories. This was different when I went to America; there were a lot of linguistics pioneers with extensive research in non-Indo-European languages.

What Professor Hill taught us in his linguistics class was interesting and capitalized on what we knew about the English language. The focus on functional linguistics also was interesting because we could apply it as trainees teaching English in high school and teacher training colleges. The basic linguistic knowledge made it easier to teach English to others and to focus on speaking, reading, and writing proficiencies. In an ideal world, this environment was conducive to teaching African linguistics, focusing on the rich contribution of languages spoken in Africa to the study of linguistics. But this was not part of the curriculum and, in fairness to Mr. Hill, he was not trained in African linguistics. African linguistics was introduced at the university by alums who had completed their master's degrees in England and had research projects on African languages.

For my first required teaching practice, I was assigned to a primary school for six weeks during summer recess. It was a blessing for those of us in education because it provided a paid summer job in the form of a living stipend. We bragged that while other students were at home waiting for the new term to begin, education students were developing teaching skills and being paid for it. We did not select the schools where we would teach; they were randomly assigned by the department. A travel allowance was made available in addition to funds to cover room and board for the six weeks. The schedule allowed enough time after the practicum to have a short recess before the new term to visit with family. Instead of going home, I decided to take a temporary job with the Ministry of Agriculture. The second and final practicum was in junior year. Teacher-education students were sent to teacher-training colleges. This was more interesting because the students were adults. I did not take a recess after this assignment either. Instead, I took a temporary job with the Ministry of Wildlife Management. It had no relation to my major, but I had friends working there who offered me the opportunity to make extra money. Working at both offices confirmed to me that I had made the right choice to be a teacher. For me, teaching was more structured and interesting than the daily office routines that my friends followed at these two ministries.

During my senior year, I started thinking about postgraduate studies. I wanted to teach at the university for an opportunity to get a master's degree as well as a Ph.D. The university offered fellowships to promising graduates who became teaching assistants. They were assigned to professors who were teaching large classes, and their main responsibility was to work with a group of no more than fifteen students from these large classes to go over the main points made in the lecture and to answer questions on behalf of their professors. They also conducted review sessions before exams and assisted professors in grading. In addition to the stipend they received for their teaching duties, the postgrad fellows had their fees paid to enroll in a master's degree. I was confident that I would be able to handle that and was determined to put in my application. During my three years at the university, I only saw two female postgraduate assistants. There were at least ten male assistants, many of them in the sciences, development studies, and history. A few were in education, languages, and linguistics. The two females were married to staff members working at the university.

At the beginning of the semester, the academic office announced an invitation to interested seniors to apply for new postdoctoral positions. I filled out my application and expressed my plan to continue with a master's degree in linguistics. Two male colleagues in my Swahili literature class also applied. Halfway through the term, I received word that I needed to go for an interview. I prepared well, thinking the interview would be mostly academic to determine my ability to serve as a teaching assistant. I knew the interviewer would be a man and at least some of the questions would be non-academic. I was right in that respect as some of the questions focused on my life plans and how long I wanted to teach before moving on. I did not understand what he had meant by move on, and when I asked for clarification, he said that he meant before I got married. He also asked whether I had someone in mind. At this point I was annoyed but tried to hide my feelings because I thought the questions were ridiculous, intrusive, and unprofessional. To stop the intrusion, I decide to give one answer to all the questions—that I had not thought about all that yet, and I was focusing on getting the best education I could. I knew that my answers were not satisfactory to the interviewer, but I was angry enough not to care. I did not hear from the academic office until the last week of the term. This was about a week after my colleagues told me that they had received letters letting them know their applications had been successful. By then, I knew I was not going to get a favorable letter, but I waited for it anyway. The contents of the letter surprised me even more. The letter thanked me for showing an interest in the postgraduate program, but they regretted that I was not a good fit. The letter added that the university was concerned that the position would be wasted because I would not be able to remain there and teach should I decide to get married because I would have to reside where my future husband worked. I felt angry and insulted by the letter, but I was not sure what part made me the angriest. I told myself that this was not the end of the world, surely there would be a million more paths to my goal. I did not know how or when, but I wanted to keep that defiant attitude. I decided not to disclose this to anyone, not even my father. I was sure he would be happy that I became a teacher after all and did not pursue my interest in architecture. He thought it was an impossible goal for a woman. I thought he was right considering what happened with my goal to work at the university.

All exams were completed by the end of March to allow the professors to grade. Then an external examiner came in to ratify the results. The ratification process included the validation of the exam questions and a thorough check to ensure the papers were graded fairly. After that, the results were announced and those who failed the final exam were given an opportunity to retake it. Those who passed were categorized depending on their score. An A+ grade qualified one to first class, A and A- were assigned an upper second class, All B's were regular second class, C and D were third class. The grades determined the salary scale, which was predetermined by the government. At that time, all the jobs were government jobs. Because the assumption was that everyone would pass the final exam, every student knew their job assignment and location before the end of the term. If a student failed both the first and second exam attempts, the government would reject them, forcing them to seek an alternative job elsewhere, an extremely difficult task. Therefore, students worked hard to avoid such a dilemma. Most students remained on campus until the results were posted on a large board near the administration block. We called that judgment day. I was thrilled when I saw my name on the list and that I had made an upper second. This meant that my teaching job was secured, and I would start at higher pay. I was ready to start packing and make the trip to my assigned post.

The graduation ceremony did not happen immediately after we finished all requirements for graduation. This was because of the way examinations were handled. Examination results had to be certified by an outside body to ensure fairness. Also, those students who had failing grades were given an opportunity to retake the exam within three months of the original testing period. Thus, a later date was selected in September of that same year, 1973, when my colleagues and I returned. Additionally, because the new academic year started in July, the university needed time to settle in before it planned for graduation. The delay was also influenced by the fact that the graduation guest of honor was the president of Tanzania. In addition to being the president, he was also the chief executive and chancellor of the university, although he was not involved in the actual administration. The administrative executives were the vice chancellor and his deputies.

All graduating students tried to attend this ceremony accompanied by family members and friends. I was fortunate because my father made the trip to the capital city, an eight-hour bus ride, to see me graduate. My brother Wenceslaus worked in the city and hosted us, as well as throwing a big graduation party for me. I had a lot of fun although I was limping after an accident the day before my graduation. He had offered to take me to the university on his motorbike for the rehearsal. I do not know how it happened, but my heel got caught in the spoke of one of the wheels and a good part of it was peeled off. He was able to stop in time when the bike started making a strange noise, only to realize that I had been hurt. Because the heel of my shoe was cut off, we stopped at a store and bought a pair of sandals for the rehearsal and graduation ceremony. After this experience, I vowed never to get on a motorbike again, and I have managed to keep that pledge. After the celebrations, I returned to my job post.

— 7 —
My First Job

My first job was in 1973 at a teacher training college in the Kilimanjaro region, two villages away from my own village and one village away from my former middle school. I was glad to be in the cold region where there were no mosquitoes and no need for fans or air conditioning. But, because it was a higher altitude, it was very cold and sometimes we got frost. This was a well-developed area in the region, one of the areas favored by the early German settlers during the colonial era. It was one of the few villages that had tarmac roads, a prestigious boarding high school for girls that was built by the Lutheran church, a couple of five-star hotels, electricity, and clean water from Mount Kilimanjaro. The hotels were an added attraction for tourists because the Kilimanjaro National Park headquarters was in this area, about ten minutes from the college. The main road led to the headquarters where tourists climbing the mountain registered before they started their trek. The Germans nicknamed this area the Swiss Alps of Tanzania because of its beauty.

The area also was famous because the municipal chief resided in this village. The Chagas who lived in the area were the only people in Tanzania who had democratically elected chiefs in the small villages that comprised the Kilimanjaro region. The municipal chief presided over the council of all these other chiefs. His powers were not hereditary like the other chiefs because his position and power were determined by the council of chiefs who could install or uninstall him. The idea was born from the need to unite the villages in the region to maintain peace and security as well as having a unified voice in local affairs before and during the colonial period.

My transition from the city back to the rural area was smooth. I did not miss the city, and I liked my new post because of its location and facilities. Families had houses and single people had apartments. My apartment was on the second floor and had two bedrooms, a living/dining room, kitchen, storage space, and a pantry. Because it was cold in this area, there was a built-in fireplace. We did not have to worry about getting firewood since it was provided by the maintenance crew who refilled it as needed. Off the kitchen was a small laundry room with a sizable sink and clothesline for drying the wash. The apartment was furnished, and all I needed to bring were linens and other personal items. The beds, couches, dining table and chairs, gas stove, and a refrigerator were already in the apartment. Coming from a dorm room at the university to a fully furnished apartment was more than I had expected. I enjoyed walking around the village, visiting the shops and open market, and hanging out at the hotels with colleagues on Sunday for a drink and snacks or sometimes a full meal for lunch or dinner if we had special guests. Being so close to home, I used some weekends to visit with my parents, leaving on Friday afternoon and returning Sunday evening. My parents were happy to have me so close to home but more elated with my career success.

The classrooms were spacious and well-furnished too. The main campus stood on a small hill on the right side of the road as one drove from town to school. The staff houses and apartments were down the hill on the opposite side of the road. Some of the staff were expatriates from England, Australia, and Germany.

I became friends with a German family and spent a lot of my time at their house learning how to bake cakes and breads from scratch. I liked making the sourdough and French breads but not the pumpernickel or rye. The color and taste of the pumpernickel bread was strange to me, and it took time to get used to it. It was easy to learn the bread-making basics because of the home economics classes I had in middle school. But my friends taught me simple methods and how to use what was available to bake the bread, such as regular cooking pots, earthen pots, or large cans that we saved after using whatever the contents were. These included lard, margarine, or other cooking oils. In return, I taught them how to make Tanzanian dishes, most of them rice or corn flour based. I was surprised that they easily adapted to the African cuisine and most of all liked goat meat dishes whether roasted, barbecued, or in a stew.

The college had a refresher course on best practices in teaching for in-service language instructors. It was organized by the Ministry of Education. A few years earlier, the government had decided to make Swahili the language of instruction for K-7. Teachers had to shift from the English-only emphasis to Swahili, increasing the number of classes taught per day and per week. Before independence, Swahili was offered once a week while English was taught in three classes a day making fifteen classes a week. Swahili grammar was taught in context since the classes focused on reading comprehension and composition. Pronunciation was accomplished in reading while grammar and morphology were dealt with in writing exercises, both at the sentence and paragraph levels. This refresher course was for language teachers who had limited Swahili classes in their educational backgrounds.

The interesting part was that the selected teachers came from different village schools in the district, including the school I had graduated from. Armed with a degree, I was automatically selected to prepare and conduct the refresher course. I had no idea who my participants would be until they arrived. They were expected to spend two weeks with us, at the end of which they would receive a certificate of completion.

On the first day of the seminars, there was an opening ceremony by the college principal. I was shocked when I entered the room and saw some of my teachers in the group. I started feeling uncomfortable with the thought of being their instructor. At this point, they did not know my involvement in the program until the principal introduced me. Those who knew me were just as shocked as I was but tried to hide it. One of them taught at the same school as my father and they were about the same age.

The introductions went well and once the principal left, it was my turn to give them the plan for the two weeks. I started by telling them that I did not know more than they did because they were the ones in the trenches and knew the learning environment for their students. I decided to use that first encounter to learn about them and their students before I introduced the new teaching concepts that would make it easier for students to learn complicated aspects of the morphology and syntax of language. The "get to know you" exercise saved the day because it was an icebreaker. By the tea break, all the participants seemed comfortable with one another. As I mingled with them during the break, I had a chance to speak to the teachers from my village. One of the teachers started bragging to the others that I was his student and then mentioned my father. I then learned that my father was supposed to be in the group but canceled at the last minute. I think my being at the school must have made it awkward for him. He knew I taught Swahili at this college and correctly suspected that I would be involved in the refresher course one way or the other. When I went home after the program, my father talked about what he had heard about the program as well as my performance as a teacher. I suspected he felt that I knew he canceled his participation, but because I did not volunteer the information, we did not bring it up. I was glad to know that my teachers from the village were proud of me and made it a point to let my father know.

I did not stay long at this college. I was transferred to a secondary school on the shores of Lake Tanganyika. To get there, I had to take two trains including an overnight one from my hometown to catch the central line. It took two days and one night on the central line train to get to the remote town of Kigoma, which sat on the beautiful lake shore. To the west was Zaire, now known as the Democratic Republic of the Congo (DRC) and northwest was Rwanda. Kigoma was so far from the east coast of Tanzania that it was easier and faster to go to the DRC and Rwanda than Dar es Salaam. It was also easier to get radio transmissions from the DRC and Rwanda than from Dar es Salaam. At that time, there was only one radio station, and it was owned by the government and called Radio Tanzania that broadcasted its programs in Swahili. The station had an English service program that was available for a couple of hours a day. Being in Kigoma felt like being cut off from the rest of the country, although it was the most beautiful place I had ever been. At night, I could see the lights in the DRC across the lake as well as the lanterns on the fishermen's boats. The fishers left in the evening, stayed out all night, and brought the fish to shore in the morning.

Fish was the cheapest meat in Kigoma. I did not know much about fish, but I had to learn how to clean and cook it. I used to chop the heads off and throw them in the bushes near my house. I did not realize the problem I was causing until my neighbors told me that an army of cats had made a home near the bush, and they found a lot of fish bones. I had to come clean and was surprised that instead of being mad at me they laughed until tears dripped down their cheeks. I did not understand what was so funny until they told me I was throwing away the best part of the fish and that the head was reserved for the head of the household or the oldest member of the family. They also told me that the head made good fish broth. I offered to save the heads for anyone who wanted them, and one of the neighbors took me up on my offer. Another neighbor told me about enhancing my chicken feed with what I was throwing away. I had a few chickens that I kept, starting with a single rooster and a hen. Before long, I had a dozen chickens kept in a shed outside my residence. I also had a small vegetable garden where I planted corn. Mixing ground corn and fish leftovers increased the nutrients I was feeding my chickens. They laid more eggs and hatched healthier chicks, but I ran into another problem.

I was so enthusiastic about making my own feed that I overfed the chickens. Consequently, they grew too fat and were unable to lay eggs. I had no idea this could happen, and the consequences were dire. I learned about it when my brother Venance who worked in public health came to visit me. We went to the chicken coop, and I told him about one chicken that looked sickly and stayed in the coop all the time because it could not run around the garden like the other chickens. I was wondering if it was safe to eat it. He went into the coop and picked it up. I saw him feeling its tummy, opened its mouth and looked in, and felt the neck area. I thought he was performing a human doctor examination on this poor chicken. After a while he turned to me and said, "The chicken has a swelling in its tummy. I think it has a lot of eggs that have failed to come out."

I was stunned and asked him what we should do. He told me to get a clean towel and denatured alcohol, clean white thread, matches, a pair of scissors, and two sewing needles with big threading eyes. Luckily, I had all these items. He told me to get the coffee table from the house and spread the towel over it. As a public health officer, he was always prepared. He traveled with a first-aid kit, which he retrieved from his big bag. He pulled out some more tools and laid them on the towel. They included what looked like a Swiss Army knife, some bandages, and medicated ointment. He also cut a branch from a tree near the house and peeled it to make thin fibers that looked like strings. I started to wonder what he was up to.

He went into the chicken coop and retrieved the chicken. He tied its feet together to restrain it and laid it on the towel. Then he put a few drops of denatured alcohol on a cotton ball and held it to the chicken's nostrils for a few minutes.

The chicken went limp and closed its eyes. I thought it was dead, but he assured me it was going to be fine in a few minutes.

I believed him because he was the one brother who had shown a keen interest in animals at an early age. No one knew how he got it or how long he had it until one evening when his efforts to hide it blew up. He knew my father did not allow dogs in the family, so he hid it as best as he could. That evening, we were sitting by the fire with my mother. Dad was away at a teacher training workshop and was not going to be home for some time. This was time to break protocols. Instead of eating at the table, we sat around the fire in the kitchen for dinner. Venance decided to share his dinner with his puppy. He wore a jacket, which made it possible for him to conceal the puppy. He took the jacket off, wrapped the puppy in it, and put half of his food inside the jacket for the dog. Everything seemed normal until one of my younger brothers saw the jacket moving. He screamed in fear, which startled everyone. Still screaming, he pointed at the jacket, which was still moving. No one was sure what to make of it, and we started to move away from where the jacket was, but Venance remained calm and seated. My mother took a long stick that was near the fireplace and lifted the jacket. We all started to laugh because out came a puppy covered with food from head to toe. Ven started to weep, picked up his puppy, and, using his jacket, tried to wipe the food off its body. My mother was sympathetic and offered to help. She took a basin, put some warm water in it, and washed the dog. She used her wrapper that she was wearing around her waist and dried the puppy, then handed it back to Venance. She also gave him another bowl of food and set some on the side for the puppy. Venance had the biggest smile on his face. Mother reminded him that dogs were not allowed, and there was no way he could hide from it from Dad. She allowed him to keep it for a while to figure out an alternative plan.

We did not know what Venance was planning to do, but the drama with the puppy was over for the time being. He left his puppy at home the next day while he went to school. His brother Cassian thought he was doing him a favor by taking good care of the puppy while he was gone. After playing with for some time, he thought since he was hungry, the puppy must be too. He got some ripe bananas for a snack and proceeded to feed the puppy as well. We do not know what happened, but the banana got stuck in the puppy's throat and it died. He hid the dead puppy from my mother until Venance came back from school. Immediately he started to look for his puppy. My mother intervened by asking Cassian what had happened to the dog because he had been with it the entire day. After a few probes, he declared that it was dead. He told Mother what had happened and showed Ven where the dog was. The banana was still stuck in the puppy's throat. Venance cried uncontrollably. Mother tried to console him and told him it had gone to heaven. Venance decided to give it a burial in the banana plantation. Those of us who were home from school offered him support by accompanying him to the burial site. Venance had a good heart and did not hold grudges against his little brother. Instead, he looked for another pet, this time a bird.

I was not surprised when he chose to work in public health, and I was not surprised that he wanted to save the chicken's life. I was confident about the plan he had in mind and was sure that the results would surprise me. Venance checked the chicken to make sure it was in a deep sleep. He untied its feet and proceeded to shear off the feathers around its tummy. Then he cleaned the area with denatured alcohol and shaved it clean. I asked him if he was going to operate on the chicken, but instead of answering, Venance looked at me and smiled. I understood to keep my mouth shut and observe. He put on some gloves and cleaned the area again with denatured alcohol. He threaded the two needles, one with the plant fiber and the other with the cotton string. I was about to witness a chicken undergoing surgery. Venance carefully cut the chicken's tummy open, pushing the organs to the side as he moved to where the eggs were piled up. I could not believe my eyes. One by one he retrieved the eggs, six total. He put them on the towel beside the chicken and cleaned up the area where they had been. I was still watching, feeling sick to my stomach, but I needed to give him moral support. Once Venance was satisfied with what he was doing, he began the process of putting the chicken back together. He moved the organs back to their respective locations and started stitching the wound. He used the plant fiber to stitch the flesh and the cotton string to stitch the tough skin. Venance cleaned the area again, put some medicated ointment on it, and dressed it with crisscrossing bandages. He tied the chicken's legs again to restrain it. He asked me to get an empty cardboard box and lay a clean towel at the bottom of it. Venance placed the chicken in it and moved it to a corner of the house. Miraculously, after patting the chicken on the back several times, it opened its eyes. I started shouting, "It's alive, it's alive, this is incredible."

Venance looked at me and smiled. He finished cleaning up and then brought the six eggs to the house. He put them in a bowl and asked me to look at them. I was shocked when he cracked the eggs; they looked hard-boiled. I was bewildered. He decided to educate me on what had happened. Venance noted that each time the eggs got stuck, the chicken's body temperature went up and the accumulated heat started to cook the eggs in the same way as boiling water. The more eggs were deposited in the area, the higher the body temperature grew around them. It must have been a while since this process started, and if it had gone on much longer, the chicken would have died because the organs in the area would have failed. I started to wonder about the human body and its processes. I also thought my brother was a genius. I told Ven I was proud of him and wondered if he would consider going back to medical school to become a doctor. He promised to think about it but added that he was happy to be a public health officer.

Venance was extremely good at his work. He became the point person for the government whenever there was an epidemic, particularly cholera. He would be sent to the affected area to provide containment measures. His colleagues praised him for his knowledge, energy, and measurable success in just a few days of his arrival. I think of him often, especially in the wake of COVID-19. He would have been on the frontline helping the country control the spread. He was working with the World Health Organization in 2007 while exploring the possibilities for career enhancement at Emory Centers for Disease Control when he got sick and died, way too young, at age fifty-three.

At Kigoma Secondary School, I taught English and Swahili to students in the eleventh and twelfth grades. Each course had three hours a week divided into grammar, composition, and literature. I also served as the language department head. It was a day school but most of the teachers resided on campus. The houses were spacious, built in the era of British colonial administration.

My house had two bedrooms, a study (office), dining/living room, kitchen, pantry, and a storage room. It also came furnished because some of the teachers were expatriates, and they were not expected to buy furniture that they would not be able to take with them when their contracts expired. These were mostly Indians and Australians with a four-year contract with renewal possibilities.

Getting this house involved a bit of drama. As a graduate, I was entitled to Grade A housing, which is what I ended up getting after threatening to reduce my teaching responsibilities. When I arrived in Kigoma, I did not get a house at school because the principal informed the education office in town that all their homes were occupied. The regional development director (RDD) decided to renovate a cottage in his compound where I could stay until they found better accommodations. These cottages were built during the colonial period as an annex where the officer could let local housekeepers or grounds maintenance personnel live. The officers lived in the mansion, which were like plantation houses in southern Georgia. Sometimes the house helper was a man with a family but most of them left their families in their home villages. After independence, the local officers inherited the facilities but used the cottages differently. Some used them as extra storage space or housed relatives who came to live with them. The RDD used his cottage for storage.

The day I arrived in Kigoma, I was utterly exhausted and happy to get to the end of the trip and settle down. It took me two nights and one day on the train to get to Kigoma.

The welcoming party at the station was the regional development director and his driver who brought us to our first destination, the RDD's house. I met his wife and children who offered a place to freshen up and breakfast. After breakfast, she offered to take me to my new home. I thought we were driving there, so I headed to the front door. She called me back and said it was not far and we could go through the kitchen. We walked toward what looked like a renovation site, making me wonder whether this was a joke. She told me that her husband had volunteered the cottage for me and that the inside was finished. She added that after cleaning the surroundings, it would look nice and be comfortable. What I saw threw out my idea of a smooth landing. I realized I had a major challenge ahead of me.

To renovate the cottage, the RDD had removed a few walls to create a kitchen and living room. When I arrived, they had just finished remodeling, but the cleanup was not completed and the debris from the demolished walls was still piled outside the door. The space created included a large room, which was designated as the bedroom with one large window and a standard bed. On the opposite side of this room, cutting across the living/dining room, was the bathroom fitted with a squat toilet and a shower. There was a small corner enclosure, which I learned later was the kitchen. They had fitted a small gas stove in this tiny space. It looked like the wall had been cut to create a small window but there was no sink or running water. The entire cottage had one water tap in the bathroom. The other tap was in the garden and was used to draw water for the plants or to wash the officer's car. In the living/dining room was a huge conference table with two office chairs. I was told that the furniture was donated by the development office and could be used as a dining table.

By the time she left, I had gotten over my shock and was angry to the point of tears, considering what I had left at my old post at the teacher's college. It did not matter to me that housing was a problem in Kigoma. At least I should have been warned instead of being assured that my accommodations already had been secured.

I returned to the main house to collect my belongings. I had shipped many of my favorite things leaving me in a dilemma because I did not know when anything would arrive in Kigoma. At the time, there was only one train a week to Kigoma. The women at the house helped me with my two suitcases and put them in the bare bedroom space. After thanking the helpers, I went into the living/dining room, sat on the chair, and started to cry. I did not know where to begin or how I was going to transition to this place. I did not have a plate, spoon, cooking utensils, or anything useful. I wished I had been smart enough to come for a visit first instead of operating on faith that all posts were equally provided for by the government. A few minutes later, one of the little girls from the main house came to the door and brought me two candles. She said her mother thought I might need them. I thanked her, and as soon as she left, it dawned on me that the cottage was not wired for electricity. I could not believe I was going to live in a cave constructed in the middle of a well-lit city. All the houses around me had electricity and I was going to use candles.

After sitting and thinking for a while, I thought I should find something to cheer me up. I decided to start clearing the debris from the front door and clean around the cottage to make it look livable. I just wanted to be outside to deflect my sad and desperate thoughts. As I was cleaning, a neighbor across the street came with her son to visit. Her husband was an important government officer who was responsible for overseeing the activities earmarked for the Kigoma region by the Ministry of Finance. Her son was a senior at Kigoma Secondary School. He wanted to meet me because I was going to be his language teacher. He was hoping that I could help him score high in his English exams so he could go to college prep school after secondary school. The mother was sympathetic to my situation. She asked her son to lend me a hand cleaning up. She went home and brought me an Aladdin lamp, an upgrade from candles. She also brought a bottle of liquid paraffin to last for a few days until I knew how to navigate the Kigoma shops and markets.

With the extra help, I was able to do more than I had expected. The entrance and the garden in front and behind the cottage looked great; my spirits lifted a bit. I thanked my newfound friend and turned to setting up the inside of the house as best I could. I had some pieces of cloth called *kitenge* that I tied together to create a curtain for the bedroom window because it faced the main road and entrance to the house. When this was done, I chose what I wanted to wear to the welcome party that had been organized for that evening. I cleaned the bathroom and shower area before I dared take a shower. I was ready by five-thirty when the driver pulled up in front of the cottage to take me to the party at the Kigoma Officer's Club.

The Kigoma Officer's Club was a gathering place on the shores of Lake Tanganyika. The government officials and businesspeople in Kigoma were members of this club and got together most evenings and weekends. It was an exclusive male club, but married women were admitted by virtue of their spouse's membership. They could also bring friends.

It was a pleasant evening with a lot of eating and drinking. The food was made at the club: barbecued goat, fish, chicken, and beef, roasted green bananas, rice pilaf, and *chapati* (Indian flatbread). I felt awkward around all these strangers, but I tried not to show it. Many of them did not have degrees and, while some respected me for my accomplishments, a few of the older men thought women did not need a degree and that I had wasted a lot of years in school. I ignored a lot of their comments and told myself that I would never see them again after that event. The party was over around eleven when the driver brought me back to the cottage. Knowing that I was going to arrive at a dark room, I had put a flashlight on the table, which I reached as soon as I entered the cottage. I first lit a candle and then the lamp. I sat on the bed for about a half-hour taking stock of the day. I promised myself to do the best I could and make an exit plan. My plan was to return to Dar es Salaam to pursue the master's degree I had been denied. I made a list of the things I thought I needed to buy and planned to go on a self-directed city tour the next day to see what I could get. By then I was tired enough to collapse.

In the morning, I took my time to get out of bed. I had brought maize flour with me, and I thought I would make porridge for breakfast. Then I realized I did not have anything to cook with or eat from. I went to the main house to borrow a cooking pot and utensils. The head of the house had gone to church, but the housekeeper was happy to lend them to me. As I was washing the utensils after breakfast, I heard a knock at the door. It was the housekeeper; she had come to retrieve the pot and utensils. I was surprised because it was less than an hour since I had borrowed them. I finished washing, then dried them, and handed them to her with a big thank you smile. At that moment, I made up my mind I was going to walk to town. It was easy to get there just following the main road, and it would take me about twenty to thirty minutes.

I was so happy to be in town and find the shops open. My biggest surprise was finding a furniture shop with nice chairs, a dining set, and things I could use to make curtains for the windows. I was glad that I had brought enough money with me because I did not know when I would be able to open an account and transfer my funds. I went on a shopping spree, making choices based on immediate needs and not duplicating too many of the items that were being shipped.

At the furniture store, I asked the owner if he knew where I could get a truck to take furniture to the cottage. He looked at me for a few seconds and said, "I think I can help you." The most surprising thing to me was that he was not going to charge me for it. On our way to the cottage, he revealed that he had seen me at the welcome party on Saturday. He told me a few things about Kigoma as we drove. He took the time to help me unload the furniture and put each chair where I wanted it. This was an unexpected gift. I had new optimism and energy as I set up the cottage to make it livable. I had to be creative because of the space limitations, but I convinced myself that I would manage until it was feasible to embark on a new plan.

I had arrived in Kigoma a week before school started to give me time to settle in. I planned to visit the school on Monday to let the principal know I had arrived. I had been given directions to the school, which was off the main road on the east side of town. Kigoma did not have public transportation. Most of the people walked or rode bicycles or motorbikes.

Early Monday morning, after breakfast, I started walking to the school. It took me twenty minutes to get there. It dawned on me that each school day would require forty minutes of walking. The sun rose early making this trip challenging because of the intense heat. On most days, by eight o'clock, the temperature ranged between 70 and 80 degrees. I convinced myself that in days to come I would get used to it

At school, I went straight to the principal's office and introduced myself. After a short talk in his office, he took me to what he referred to as the staff room. This was a big conference-like room with desks and chairs. Behind the chairs was a cabinet with shelves that held books, notebooks, maps, and other teaching supplies. Each teacher had one of these shelves but used them differently. There was no one in the staff room at the time; the teachers were still on Christmas recess before the start of the new school year. The principal showed me my desk and then led me on a tour of the school grounds. We toured the classrooms and then the staff quarters. Before we reached the big houses that looked well-built and cared for, he commented on the generosity of the regional development officer who gave up his cottage so I could use it. I did not know whether he had anything to do with it, but as the principal, he must have been informed about my arrival and the house assignment process. All government sectors, including education, were under the RDD's office, which oversaw the administrative aspects of government business in this region. I decided not to comment until the tour was over. Walking back to his office, I asked the principal how one qualified to get a house on the school compound. He told me every teacher was entitled to a house and that I was entitled to a Grade a house because of my education level and the position I was going to hold at the school. He quickly added that he could have assigned me to one of the big houses occupied by single male teachers and they could share a house. I asked him what had stopped him from doing that. He confessed that it would be hard to convince them to share accommodations. I jumped at the opportunity to tell him about the conditions of the cottage I lived in, inadequate running water, no electricity, and the walking distance to and from school every day. I also told the principal it was awkward that some of my students were coming out of mansions occupied by their parents who worked for the government in the city, and I was coming out of a cottage referred to as "the servant's quarter." He kept quiet for a while and then said he was sorry and that it was all he could do at the time. I did not say much more and pretended I was not annoyed by his explanation.

Back at his office, the principal laid out the duties of the language section head, handed me copies of the books for my classes, and showed me where students obtained their supplies. Fortunately, there was someone assigned to handle the students, and I just had to worry about managing the unit and teaching the assigned English and Swahili classes. We came to the end of the visit, and I thanked the principal for his time. I noted that he might see me again that week as I planned to return to set up my desk and create some lesson plans.

On my way back to the cottage, I found myself getting more and more disturbed by my moving to Kigoma. I started planning for an early exit. I was determined to apply for postgraduate studies as soon as possible. I knew there was no reason I would not be accepted especially since I would be going in under the Ministry of Education staff development program rather than the university's postgraduate fellowship program. I also planned to meet with the RDD to request that they wire the cottage for electricity, put in a kitchen sink, and connect a water supply line from the same source as the tap in the garden. In that way, I would have a second tap in the house and would not depend on the one in the bathroom. I was skeptical about these demands, but I was willing to give the plan a shot using charm rather than anger.

With all these thoughts going through my mind, I forgot about the scorching sun and the distance to the cottage. I found myself at the entrance, thoroughly wet with sweat. I took a shower and had a snack. I wanted to have a head start on the literature books I was going to teach and the lesson plans. I also had some more creative work to do to make the cottage welcoming. I measured the windows and planned a trip to town the next day to get the curtains made to fit.

Around four-thirty, I saw the RDD's Land Rover entering the compound. I gave him about half an hour and then went to the kitchen door to request a talk with him. His wife worked at the bank in town, and she did not come home until late in the evening. The housekeeper went to relay my request. The RDD came out to meet me and stood on the top stair outside the kitchen. The stairs to the house, both at the entrance and to the kitchen, are steep. While he was standing there, he looked like he was on stage, and I was in the audience at a theater. All the houses in this subdivision had a raised foundation, perhaps because they were built on a hill with a steep slope overlooking the lake. The incline provided a spectacular view of the lake, especially at night with the fishing boats and a panoramic view of the lights in the Democratic Republic of the Congo on the western shore of the lake.

The RDD must have been in the sitting room reading the daily newspaper when I asked for him because when he came out, he was still holding the paper in his hand.

"Welcome, how are you settling in?" he asked.

"Very well, thank you. I am trying to manage; it is a big transition from Kilimanjaro to Kigoma."

"I understand," he said.

"You wanted to speak to me," he asked.

"Yes, I have a couple of requests to make. I was wondering if we could extend electricity to the cottage and add a sink in the kitchen. We can easily connect a water line to the source of the tap in the garden," I replied.

"Of course, that is simple. I will send the engineer from the office tomorrow morning. By the evening this should be done. I will also call the power company to authorize extending the power line from the main house to the cottage. I do not know why we did not think of this earlier. Certainly, this will be done by the end of tomorrow.

I thanked him and started walking back to the cottage. He called back, "Let me know if there is anything else you need, anything at all."

"Thank you," I replied.

Back in the cottage, I could not believe my skepticism was unfounded. I wondered whether this was an advantage to the owner rather than help for me. I was providing a legitimate excuse to remodel this cottage using government resources. I knew that when I moved out, he could use it for anything he wanted, and it would be the best extension of its kind for the main house and different from the other cottages around town. It did not matter to me at this stage; I was trying to survive the challenges at hand.

As promised, on Tuesday morning, a truck pulled up in front of the cottage at nine-thirty in the morning. The gentleman introduced himself as Peter and added that he had been sent by the RDD to supervise electrical wiring and the installation of a sink in the kitchen. He had an electrician with him and a plumber. They came inside the cottage, did a brief survey, and then got to work. They asked me where I wanted the lights, sockets, and switches. Using my experience from the other places I had stayed, I selected at least four sockets, two in the bedroom and two in the living room. I did not need one in the kitchen because the space was too small, and I already had a gas stove in it. I chose two light switches in the bedroom, three in the living/dining room, and one at the entrance to the bathroom. As for lights, I had two fluorescent tube lights in the bedroom, two in the living/dining room, two regular lightbulbs in the bathroom, one in the kitchen, and an outside light to illuminate the entrance. Despite its limitations, this little cottage was becoming livable. I was beginning to like it, but these additions did not change my feelings overall.

I left the men working and walked to town to get my curtains made. I used the time to continue exploring. By the time I returned, about four hours later, the electrician had finished wiring the cottage and was waiting for the utility company to complete the connection to the main house. They told me this was a half-hour job. The plumber already had installed the sink and was working on connecting the waterline. The sink transformed the kitchen. I was so happy that I did not have to use a bucket to wash dishes or depend on getting water from the garden or bathroom taps. The entire operation was completed by two in the afternoon. I was so happy that I did not have to use the lamp or the candles that night—or ever—unless there was a sudden power interruption. The cottage was transformed into a little resort. I was now ready to meet my students and teach.

The first day of school was interesting. I had spent a year at a teacher training college, but now I had to adapt to teaching eleventh and twelfth graders. For my major training, I had selected teacher education to avoid teaching secondary school, but I could not run away from it. As a government employee, one did not choose an assignment. It was determined based on the government's needs. The thought of preparing the twelfth graders for their national exit exams was daunting. In this culture, students' successes or failures are attributed to the teacher. I was not prepared to take the blame if the students did not pass their English and Swahili exit exams. I was confident that I could have success with the eleventh graders because they had two years with me. I was not sure about the twelfth graders because I did not teach them the previous year. If there were weaknesses, it would be impossible to bring them up to speed in nine months before they started their exams.

I tried to settle into the job, but a month later, I started getting weary of the walk to and from school five days a week. I decided to talk to the principal about using the information he had provided on the first day I visited with him. I proposed that he should use his position as the principal to convince the two teachers to share a house so I could get a house at school. I made it clear to him that being the head of the section and teaching the upper classes was taking a toll on me when I had to spend forty minutes walking to and from school each day, five days a week. I decided to give him an ultimatum. If by the end of the week I did not get accommodations on campus, I would reduce my teaching load to just the twelfth-grade classes and heading the unit. The principal thought I was kidding. I thought I would hear something by Friday, but nothing happened.

On Monday of the following week, I arrived at school as usual. I taught my first class for the twelfth grade, English Three, which was literature. After that, I had Swahili with the eleventh graders. I decided to go to the office rather than class and graded homework submitted by the eleventh graders. Ten minutes later, the second head came to the office and asked why I was not in class and that the students were noisy and disturbing the other classes around them. I politely told him I had informed the principal that I was not teaching that class anymore and that he should find a substitute. He left to see the head teacher. He did not come back to the staff room. I taught my other twelfth-grade classes that day and avoided the eleventh grade. Around two in the afternoon, the head teacher sent for me. I went to his office thinking he was going to threaten me or tell me that he had reported the matter to the Ministry of Education and that I had refused to teach. Instead, he told me that he had allocated housing on campus for me and that the painting crew would start the next day to make it ready to move in by Friday. I could not believe my ears. I tried to hide my excitement but assured him that I would resume my normal teaching the next day. He seemed happy to hear that. I returned to the office, finished what I was doing, and left campus excited by the turn of events. This was another surprise because I was not sure how my strike was going to go down. I did not know whether the Ministry of Education had been informed of my strike and whether they would have sided with me or the school since I was so new. I found myself saying, "Ask and you shall be given." Indeed, I took a chance, and it had paid off a second time since I had arrived in Kigoma. When I got back to the cottage, I decided to go to the main house to let the RDD know that the school had allocated a house for me on campus. With excitement, he asked me when I planned to move. I did not know because it required securing a truck to move everything I had accumulated in the cottage. He volunteered to send a truck from his office and a couple of people to help me move when I was ready.

On Thursday before school was over, the principal took me to inspect the painting and cleanup. I was satisfied, and he officially gave me the house keys. That afternoon I went back to the cottage and told the RDD I wanted to move on Friday. Around three in the afternoon, the truck pulled up at the cottage, and the crew started loading it with my belongings. After the delivery, I returned to the cottage. I spent the last night there cleaning up.

Early Saturday morning, the RDD's driver gave me a ride to campus with the rest of my belongings. I was now officially an on-campus resident. It was refreshing to have the additional space. The two furnished bedrooms, a study/office, living/dining room, kitchen, pantry, and storage space were a welcome change from the cottage.

Both of these big houses were on a hill and had flower beds with a variety of weather-resistant plants. There also was a sizable garden space for those who liked to grow their own vegetables. I loved the great view of the lake from my front porch where I could sit in the evening to watch the fishing boats and admire the cluster of illuminated buildings in our neighboring country, the Democratic Republic of the Congo. From campus, a trip to the lake was about ten minutes. Many of my colleagues went to the lake early on Saturday mornings to gather sardines that had been washed ashore by waves during the night. Villagers collected the sardines in buckets to sell to others or for personal use. I made a trip once to get sardines that I mixed with cereal for chickens.

With the immense space, it was easy to set up my new residence. The neighbors, an Australian couple and a family from India with two children, came to welcome me and asked if I needed help. The Australian family brought me homemade scones and the family from India brought flatbread. It was indeed a very warm welcome. The little girls from the Indian family loved my home and often would sneak out of their house through the back door to mine. The first time they did that, I was not sure what to do with them. In Africa, kids can go from house to house to look for playmates. Parents were never concerned. I decided to walk the children back home but said I would ask their parents, and if they allowed them, the girls could come visit anytime I was home, but they had to tell their parents first. When the parents saw me walking them back, they apologized if their children had disturbed me. I seized this opportunity to assure them that it was fine for the kids to come anytime, and I was more than happy to have them. That established a lasting friendship with the parents, which allowed us to sit together on the porch many evenings enjoying the view of the lake. The children called me Auntie and cried whenever I told them I was traveling and would be away for a few days. The children forced us to become an instant family from two different cultures. When I started raising chickens, they helped me collect the eggs. They liked the idea of making a tray to take home. They also got chicken to eat, which their garden helper killed and cleaned for them because they did not know how to do it. They liked chicken curry, which they always shared with me.

After the stress created by the accommodation challenges was lifted, teaching at the school became fun. I had more time with the students in different school clubs. I encouraged them to start a cooking club. We met on Saturday at my house and made a lot of African doughnuts (*mandazi*), tiny baking soda cakes (*hafkeki*), and flatbread. The students sold them to their friends on campus and banked the proceeds. The club was well organized and had a president, vice president, treasurer, and secretary to take minutes whenever they had meetings. I loaned them the initial capital, which was paid back in a month. They waited until the end of the term to divide the shares. They were so successful that they continued the club for a year after I left the school. They recruited members from all the classes to ensure sustainability. I was proud of them and told them they needed to keep their entrepreneurship going after graduating from Kigoma.

In addition to teaching at Kigoma, the Ministry of Education gave me the responsibility of conducting English proficiency tests for secondary schools in the Kigoma and Tabora regions. This was a required test for twelfth graders before their final written English exam. The results were sent to the ministry, and they were added to the final grade. The required proficiency level was three, a high intermediate level of English. None were expected to attain level four, an advanced level of English, or level five, which was considered proficient in English, a level given to those who speak English as their first language. These standards were established by the colonial administration and were kept in the system for years after independence. When Swahili attained an equal emphasis in the country, the need to conduct the English proficiency test was eliminated.

I enjoyed this responsibility, which took place between July and August each year. The government paid for my transportation and a hotel room when I traveled out of town. I did not perform language-testing of my own students; another tester was sent to do that. At the schools, it was the responsibility of the head teacher to provide the testing rooms and to organize the testing schedules. It took at least two days at each school. While I enjoyed doing it, it was sad to watch the students stressing over this. My job was to calm them down to get them to a point where we could have a conversation without thinking too much about the exam. The more I did this job, the better I became at managing the environment to reduce the students' stress. It allowed me to test objectively and provide a realistic grade that was not affected by the students' fears.

I stayed at the school for two years. My departure was another outcome of my persistence. Once settled in my teaching, I decided to apply for postgraduate studies to get a master's degree in linguistics. I applied in January, and I did not hear from the university until April. When the letter arrived, I did not expect good news. The letter congratulated me for being accepted and informed me that the degree program was by thesis. This meant that I did not need in-class instruction but would work with my committee, selected by my major professor who was appointed by the university. The committee would work with me to select the area of research and through advisement select the appropriate literature that would inform and support my research. As such, I did not have to quit my job to do it, but it would be advisable to be within reach for meetings and consultations with my committee, three of whom were on the campus of the University of Dar es Salaam and one in Uganda and who also served as the external examiner. With these instructions, I had to seek a transfer from Kigoma to Dar es Salaam. I knew the transfer request was going to be an uphill battle because the government did not grant transfers easily. Filling positions in the remote parts of central, western, and southern Tanzania was hard and once filled, the government liked to forget about it and let those serving be there in perpetuity. I knew I could not resign from my teaching job because getting another job would be hard. Nevertheless, I wrote a letter to the director of secondary education at the Ministry of Education requesting the transfer.

Within two weeks, I received a response. As expected, the answer was "No, not at this time." The reason given was that it would tough to find a replacement on such short notice. He indicated that he would let me know as soon as he found somebody. I felt crushed and that this was unfair and the director had no right to deny my request for further studies. I had to put on my thinking cap to look for a plan that would get me out of this quagmire. I decided to write a letter to the university registrar to let him know what the Ministry of Education had said. Like me, the registrar was a native of Kilimanjaro, and the Chagas are known to fight hard for education. He sympathized and promised to find a way to help. I did not know what he was going to do or if he would succeed. Three weeks later, I received another letter from the director of secondary education informing me that I had been granted a transfer to the Dar es Salaam

Secretarial College to teach communication skills. When the letter arrived, I was in the office. After reading it, I started jumping up and down in a happy dance. My colleagues were confused; they could not understand what this was all about. After I calmed down, I announced I was moving to Dar es Salaam to teach at the government's secretarial college.

News of my departure was not well received by my students, particularly the cooking club. I wanted to ask one of the teachers to be their adviser but the club members were apprehensive. Then I had the idea of finding space for them in the kitchen to continue making the snacks. The chef was agreeable to the plan, and students were confident they would be able to manage without me being at the school. After I left, I kept up with them for about a year and later learned they had decided to let it go. What I was sure of was that losing the club was minor compared to the experience that would live with them forever.

The reporting date at my new job assignment was at the end of the term giving me a two-month window to get ready to move. I started to organize the packing and shipping of what I wanted to take to Dar es Salaam. Being so close to the railway station was an advantage. I found a shipping company that came to the house and packed everything I wanted to send in a specially designed crate. Once that was gone, I was left with suitcases of clothing and other small items.

The school had a farewell party for me, and my students gave me a memorable gift—an elaborately embroidered bedspread. They also came to the train station to see me off. It was a sad day saying good-bye to my newfound family in Kigoma. I was ambivalent about returning to Dar es Salaam because I did not like living in big cities. Apart from being more expensive than Kigoma, Dar es Salaam was congested with people and cars. The buses were always crowded, noisy, and hot. I did not know what part of the city I would find a place to live nor how difficult it would be to get to and from work. Nevertheless, I was happy to go back to the big city to realize my plan for an advanced degree.

I was required to finish my master's in linguistics in two years, one year of research and advisement and one year of writing. Teaching at the secretarial college made it easier to do this. The schedule was not as rigorous as at the secondary school, and I was teaching adults in the workforce. I had to shift gears and prepare to teach text analysis, principles of public speaking, reading/comprehension, and writing. I had some flexibility too and was not required to be at the college in-person unless there was a meeting or classes. I taught one class three times a week—Monday, Wednesday, and Thursday. This allowed me to devote Tuesdays and Fridays to research and writing. My adviser had set meetings twice a month, saving me frequent travels to the university, which was a one-hour bus ride each way.

For my thesis, I chose to research my first language, Chaga, to explore cohesion in discourse texts, a concept discussed by Halliday and Hassan in their celebrated book "Cohesion in English" (1976).[3] According to these two linguists, cohesion is a semantic concept that deals with the relations of meaning in a text. Basically, it is the grammatical and lexical linking within text that holds it together and gives

---

3 cf. Cohesion in English (1976) by M.A.K Halliday, Professor of Linguistics, University of Sydney and Ruqaiya Hasan, Associate Professor of English and Linguistics, Macquarie University. Publisher: Routledge, Taylor and Francis Group, London, and New York.

it meaning. This was my introduction to discourse analysis, an interest I developed when I started teaching communication skills at the secretarial college. I chose Chaga because this was the first time a linguistic description of this language would be done. It was an opportunity to explore a language that was near extinction as Swahili dominated the medium of communication across the country. I also wanted to write the thesis in Swahili. My supervisor was surprised, but he did not discourage me when I shared my plans with him, though he noted it was going to be a difficult task. I was the first to write a thesis or dissertation in Swahili, and about the Chaga language. He was concerned that the limitations of Swahili would not allow me to capture the full extent of the concepts I planned to explore. I convinced him that I was limiting my exploration to word order and contextual meaning. I was sure I could find or coin terms that fully exploited the themes for the intended thesis. We agreed on the game plan, that I would select the concepts I wanted to work on and then share the translations with him. Once approved, I started to examine the concepts in Chaga using Halliday and Hassan's theory on cohesion in English. I wanted to find out how universal their theory was or whether it was language specific, in this case, English.

Two years passed quickly, and before I knew it, my thesis was reviewed and approved for defense, which was scheduled in July to allow time for changes and submission before graduation in October. The fourth supervisor was also the external examiner and received a copy ahead of the defense date. I had no experience in this process, but my supervisor was confident that all would be well. Because this supervisor was in the driver's seat, he told me he planned for a short defense.

On the day I was to defend my thesis, I arrived at the university and went straight to my supervisor's office. Because he was in class, his secretary gave me the room number where the defense was to take place. I headed there and waited for the committee to arrive. When all were present, my supervisor introduced the subject matter and asked me to summarize my findings. I thought I was going to field a lot of questions, but each member took turns talking about what they had learned from my findings. They were all pleased with my work and congratulated my supervisor on taking a leap of faith to direct the first thesis in Swahili. The meeting took about thirty-five minutes and ended with the external examiner congratulating me on a job well-done. She told me to go celebrate and wait for graduation in October. My supervisor told me to check with him in a month to get a copy of my thesis. It was the supervisor's responsibility to submit the thesis to the library where copies were made and bound. A copy was reserved for the library and others were sent to the supervisor who distributed them to the committee.

This was it. I had finally gotten what I was denied four years earlier to the day. I also was glad that I accomplished this before those selected for postgraduate education began their studies. Now I could go back to teaching and request a salary raise because I completed staff development.

— 8 —
New Opportunities

In the Bible, Psalm 118:22 says: "The stone the builders rejected has become the capstone."

The first time I read that was in high school Bible class and I linked it to the story I told earlier about the girl who disparaged a pebble, saying it was ugly and threw it way, despite her friend's assertion that it was beautiful. Later they found the pebble had turned into a rock, blocking their path home. Unashamedly, I likened it to my return to Tanzania's capital city to finally pursue the master's degree I was denied earlier. Furthermore, after completing my degree, I never imagined that the university would contact me with an offer to join their research team at the then Institute of Kiswahili Research now known as Institute of Kiswahili Studies. Two of my colleagues who were successful in getting the post-graduate fellowship I had been denied worked at the same institute. I thought it would be interesting to join them with a master's degree in hand while they were still working on theirs four years later.

About a month after successfully defending my thesis. I received a phone call at work asking me to come to the university for a discussion. I was not sure what they meant, but out of curiosity, I decided to go to meet with the Institute's director He welcomed me warmly and said he was looking to revamp the institute's research group and was told by my major professor that I was a good candidate. I told him I was flattered and would give it some serious thought and would get back to him in a week.

Riding the bus back to my office, I kept thinking about the offer. The salary was significantly higher than what I was earning at the secretarial college, but what excited me most was that I did not apply for the job; I was being sought out by the same institution that earlier had rejected me. I decided even before I got back into town that I was going to accept the offer for two reasons. One, it would be an opportunity to begin pursuing a Ph.D., and two, I was getting what I had been denied four years ago, but this time an application was unnecessary. I decided to wait out the week before telling the director so as not to appear too eager.

Instead of calling at the end of the week, I waited until Monday and planned to take the bus to the university to give the director my answer in person. But before leaving, I received a call from the office of the president of Public Service Management and Good Governance to meet with the director there. Because the secretarial college was under the jurisdiction of this office, I started to wonder what this was all about.

When I arrived at the director's office, he was sitting behind an executive desk with three phones on it. It was intimidating—more so because he did not rise to shake my hand. I had to stretch mine across to reach his. He went straight to the topic at hand, which concerned a new assignment for me. He wanted to know whether I would be interested in a position they were creating to oversee staff development training. My responsibility would be to develop the programs and create schedules for the different departments. He added that I would have a support staff of three and the job might require frequent travel. The job description sounded interesting but the idea of pushing paperwork was not appealing. I thanked him for the offer and promised to decide in a few days. He did not look happy that I did not jump at the opportunity. I also was not sure I could work with him as my boss. I left and returned to my office. I could not stop thinking about the encounter but did not talk about the offer because I did not want my decision to be influenced by anyone. By the next day, I had made up my mind that a university job was far more liberating than a government post. I had developed a liking for institutional settings and was not ready to give that up and join a bureaucracy that required intense public relations negotiations. At the office that morning, I debated whether I should go in person to announce my decision or write a letter. I decided to write a letter and hand deliver it to his secretary. A week later, an announcement was made about the new position and the name of the person who would head it. Meanwhile, I made my scheduled visit to the university to accept the research position. I had two months to finish the teaching term at the secretarial college. When the time came, I announced my departure and moved on to the university, where I had landed when I first came to the city.

My first day working at the campus, nicknamed the Hill, was October 1, 1997. It was a major transition from teaching to exclusively doing research. I was happy to return to a rigorous academic environment, debating colleagues about linguistics and ideas and concepts in African languages spoken in Tanzania.

There was some flexibility on research focus, but the director wanted me to work on nominals in the Swahili language. I was supposed to examine earlier work done by. Joan Maw, an English linguist who had spent substantial time in Tanzania researching Swahili and had published a book on some aspects of nominals in the language. The director wanted me to explore the concepts I had focused on in my master's thesis. I had the liberty to do a comparative study. That charge seemed doable because the area of research was still fresh in my mind. I also decided I would combine desktop and field research. Being at a research center was an advantage too because I could bounce ideas and share my findings with colleagues for comment and discussion.

I settled into my new job quickly, partly because I knew most of my colleagues at the department. The year ended and the new year began with plenty of work to do. I did not miss teaching. I enjoyed the ability to set my schedule and be my own boss except when I had to participate in department activities. The department assigned two students to help with data collection, specifically in the field. This made my work simpler, allowing me to process the data rather than collect it. By mid-July 1998, I had accumulated a lot of data and was making reasonable progress analyzing it. But this was disrupted by an unexpected change of plans, a chance to experience America for the first time.

The Peace Corps office in Tanzania offered me an opportunity to travel to the School for International Training in Vermont, to train American Peace Corps volunteers for deployment to Tanzania. How I was selected to participate in this program remains a mystery because I did not know anyone working with the American Embassy in Tanzania or with the Peace Corps. However, I knew of the Peace Corps program because when I was in high school, there were a lot of corps volunteers teaching at different schools, including Korogwe Secondary School. It was there I met Miss Little, a Peace Corps volunteer who taught math at the school. It was now ten years later and the Peace Corps was knocking at my door at the University of Dar es Salaam. A young woman, whom I got to know later as Ann, had been sent to the university to recruit two individuals to travel to the United States with her for eight weeks to train new volunteers for Tanzania.

Ann explained why she had come to see me. Her introduction was in Swahili, something that took me by surprise because she was exceptionally good—at the native speaker proficiency level. She then explained that she had come to interview me for a potential project that involved preparing Peace Corps volunteers from the United States to go to Tanzania. The United States was reopening the Peace Corps program there after suspending it for a few years. My role was to participate in required pre-entry language and culture training for the volunteers, along with creating a manual. This opportunity was to take me out of Tanzania for the first time—to the Western Hemisphere, the United States of America.

Ann wanted to recruit two women in Tanzania and my department suggested me, which explained her unannounced visit to my office. She had already met with another colleague in the department, Sitna Masamba, who had accepted the invitation. Although it was a daunting task, I did not think I should pass up the opportunity. I accepted the offer, and Ann preceded to set up a meeting date to discuss the plan in detail. She left my office, leaving me to process what had just happened. I could not believe it was real. It was an opportunity I had never dreamed of.

That was Tuesday. Ann returned to the university on Friday morning to meet Salima and me to begin planning. A driver from the Peace Corps office took us to the hotel where Ann was staying. We spent the afternoon poring over the teaching manual she had prepared. For someone who had learned Swahili in the United States, she was exceptional. We ended our day around four in the afternoon with a plan to meet again on Saturday.

On Saturday, we spent the morning editing the manual and the afternoon on orientation about what lay ahead. Although the orientation was long and tiresome, we were excited about the project though a little fearful about heading into an unknown world and facing the daunting task of training the volunteers in language and culture in just eight weeks.

At the end of the day, Sitna and I were given four-hundred Tanzanian shillings (about $50 at the time). This was a lot of money for a day's work when compared to my full-time monthly salary, about $85, at the university. We were told that the $50 was compensation for the time spent reviewing the manual and getting oriented. Sitna and I were amused by the gesture because it was beyond our wildest imaginations. Although we did not show any emotion about this generous gift while in Ann's presence, we could not contain ourselves once she left the room. We put the money on the table and stared at it in disbelief. We started wondering about the trip, the new land we were about to visit, and the people we might meet. I thought about John Steinbeck's novel, *The Grapes of Wrath*, particularly the infamous line, "In California, we eat what we can and can what we can't." Indeed, the land of plenty. We wondered what that land looked like and how we would feel when we saw it, let alone lived there for eight weeks.

When Ann came back into the room, she handed us letters for the Ministry of Home Affairs office to process our passports. We left and headed home, planning to meet the following Monday at the Ministry of Home Affairs to process our passports.

To our surprise, the passports did not take long. We received them in just two days. That was Wednesday. We surrendered our passports to the Peace Corps representative to process our entry visas to the United States. To get a visa was a much simpler process compared to now. By that Friday, we received our passports with a visa that allowed us to enter and stay in the United States. Ann wanted us to meet with her the following week for the final orientation before we left that weekend for America. The meeting was on Friday when she handed us our tickets. With the tickets, passport, and visa in hand, reality started to set in. We were on our way to a place that seemed so far away and foreign. I felt a mixture of excitement and fear. It was happening too fast. In a little over a week, starting from Ann's visit to my office, we now were ready to fly to America. Even my family did not believe it. They thought I was pulling their legs until I showed them my passport and a travel visa. Sitna and I made an appointment with a hairstylist on the Saturday before our departure. We had agreed on our outfit and a uniquely African hairstyle. We chose a style called "queen's crown." The dress had a maxi skirt with matching top made from local fabric called *kitenge*. It took about eight hours to get our hair done. We looked in the mirror and agreed it would make an impression when we arrived in America. At the airport on the day of departure, everyone was looking at us as we made our way to the gate. We looked stunningly gorgeous, like real queens of Africa, elegant, walking majestically onto the plane. This was the beginning of experiencing America.

Our destination was Brattleboro, a small town in Vermont where the School for International Training (SIT) was located. This was to become our home for eight weeks. We started to experience culture shock upon arrival at the airport. It was the first time we saw escalators. To hide our fear of getting on these moving stairs, we just laughed and laughed. We gathered our long, flowing skirts to avoid tripping. My colleague was more courageous than I was; she attempted to step on the escalator. As she did so, she looked like she was going to fall. A good Samaritan stepped in, took her by the hand, and helped her to the bottom of the escalator. This paved the way for me. Another sympathizer helped me down to the bottom. We got off and stood for a while just laughing until tears dripped down our cheeks. We were trying to hide our embarrassment.

At SIT, Sitna and I joined four other Tanzanians who were already in the United States studying and working at different universities such as Cornell, Rutgers, University of Massachusetts, and University of California, Los Angeles. The training team consisted of seven people including Sitna and I. Our responsibility was to teach the Peace Corps volunteers the Tanzanian history, the Swahili language, and Tanzania culture. Our interaction with the Peace Corps volunteers was intended to create a soft immersion environment to learn language and culture through daily participation. Our task was to facilitate their understanding and preparedness to adapt the culture and working environment for their two-years' service in Tanzania.

There was a lot to learn on this first trip. The culture shock alone was tremendous, starting with the food, the people, the climate, and the task of teaching Americans. I had some experience teaching Swahili to foreigners. I had worked with the Danish and Swedish Embassies in Tanzania, teaching field officers the language and culture of the country. But teaching Americans was different, especially because we had a set curriculum that we were to use. Before we started training the Peace Corps volunteers, we were given an orientation on the principals of second language teaching. We did not think the orientation was helpful because the organizers used Spanish to illustrate instead of Swahili in training us how to teach Swahili to the volunteers. I wondered why Ann did not use Swahili considering that she spoke Swahili fluently. It became hard for us to adhere to the methodology, which made the whole exercise seem superficial to us. It would have been better to demonstrate the principles by teaching an actual class, videotape it, and then use it as the basis for the orientation workshop, looking at what worked and what didn't. I was puzzled by the fact that they did not consider the experiences of the four Tanzanian colleagues we met at SIT, recruited from different institutions in the United States, and who were seasoned Swahili teachers in the United States. Sitna and I overcame our shock and adapted to the "new ways of doing things" by applying the principles, the best we could, to this specific task of teaching the Peace Corps volunteers.

I mention food as a culture shock because the menu was foreign to me. I came from a culture where everything called food was a steaming hot dish. To start with, my friend and I thought it was a joke to brew tea from a tea bag in a cup and then add cold milk to it. Tea, in our mind, was cooked. The four Tanzanian colleagues we met at SIT understood this tea culture because they had been in the

United States for several years, as students and now professors at different institutions. However, they confessed that at their homes in the United States, they still "cooked" their tea.

How do you cook tea?

For black tea, one boils water, and then adds the tea leaves. For tea with milk, which we referred to as colored tea, equal parts of water and milk are mixed and brought to a boil. Then, loose tea leaves, not a teabag, are added, and let to simmer for two to three minutes while stirring to prevent it from boiling over. If busy, you can put a clean saucer in the pot with the tea (this trick works for milk too). This will prevent it from frothing and boiling over. After that, the tea can be poured through a strainer into a teapot or strained directly into a cup.

The description is a modified version of English tea culture where hot water is poured over loose tea leaves in a teapot and left to steep for a few minutes. Then the tea is poured into a cup and either hot milk is added, or it is left black. In my culture, the English style produces tea that is weak, while the American use of teabags produces even weaker tea because it is not perfectly cooked. The English method would be a welcome alternative to the use of a teabag. At Brattleboro we did not have a choice. We knew we would have to endure it for eight weeks. My friend and I nicknamed it "fake" tea. Now, after living in America, I do not consider it that anymore. I adapted, and even though I have the option and ability to cook tea, convenience wins the day.

Another culture shock was sandwiches. This was the first time I ate a sandwich. It tasted fine although it was served cold. For the first couple of days, Sitna and I could not finish a whole sandwich. The other food shock was rice.

Rice is part of African culture, and it is cooked in different ways depending on what part of Africa one is from. In East Africa, adaptation has been key since rice is not a staple. It was introduced to Tanzania from Asia. It can be served as plain white rice, *pilau* (cooked with different spices with or without meat and occasionally adding a few potatoes), *biryani* (spiced vegetarian rice), *mseto* (rice mixed with beans), and many other creative ways to make it an interesting and appetizing dish. The first-time rice was served at SIT, it was from a tin tray sitting on a warmer. The rice was swimming in water (my description) and served using a slotted spoon. We nicknamed the dish "swimming rice." In our opinion, the rice was half-cooked.

We did have an opportunity to share Tanzanian food culture at SIT. We convinced the chef that it was important for us to show the Peace Corps volunteers our food customs since they were going to live in Tanzania for two years. We were granted a day in the kitchen with the crew and a few volunteers. This went very well, and we were surprised when we saw some of the dishes we introduced show up occasionally. The biggest influence was on the rice. We never had rice swimming in water for the remaining time at SIT. I am sure it was a good cooking experience for the kitchen staff as well as the Peace Corps volunteers.

The other culture shock was more in interpersonal relations. Because Tanzania did not condone racism and discrimination, it was not always obvious to Sitna and me that some comments, actions, and reactions had racial undertones. The other Tanzanians who had been in the United States for a while could spot signs of bias.

When we arrived, all of us (trainers) were assigned rooms in the same dormitory as the trainees. Sitna and I did not think much of it because we had no expectations. However, our US-based colleagues were not happy because some of them were adults with families of their own. It was difficult for them to be in a dormitory where bathrooms had limited sinks, toilets, and shower stalls. This meant one had to leave his or her shared room, go to the bathroom area, and wait for a free toilet, sink, or shower. The trainees were much younger than us, mostly recent first-degree graduates.

About three weeks into the program, signs of discontent about the accommodations started to surface. We had a group meeting one evening and it was decided that we should discuss this issue with the Peace Corps representative on campus who was our immediate supervisor. Ann, the recruiter we had met in Tanzania, traveled a lot between Brattleboro and Washington, DC, where the Peace Corps had its headquarters. She was not at SIT for the most part, and when she was, she stayed at a hotel downtown with the other program representatives. The hotel was within walking distance of the campus.

Once we agreed that we should have a meeting with the supervisor, we laid out some strategies to safeguard relations and maximize our chances of a good outcome. We sent a two-person delegation to set up the meeting with the supervisor. We agreed it should be the next day after classes. Our strategy was to have two individuals present the issues to the program supervisor, specifically our concerns about the accommodation situation and about having a more defined schedule that allowed personal time.

The meeting did not go as we had hoped. We had intended to forge cross-cultural understanding, emphasized in our training of the volunteers. We felt there was a lack of understanding of our culture by our host/supervisors.

In the meeting, the designated speakers took turns explaining the two main issues as best as they could. After they finished, the supervisor paused and then spoke. Concerning accommodations, she said the goal was for us to be in the same living quarters to maximize the trainees' knowledge of language and culture through soft immersion and to enhance their verbal proficiency. She added that our work hours should be considered 24/7 for those very reasons. Individually, we felt defeated. But what she said next took us all by surprise. She noted that we were lucky to have this opportunity to be at SIT because the Peace Corps was paying a lot of money for the program. We should be grateful because we could not make that kind of money in Tanzania even if we worked for ten years. Then she added that the living quarters were far better than a thatched hut in the middle of Dodoma, implying that we were coming from extremely poor conditions and should accept anything that was an upgrade. We looked at one another and I could see the disbelief in our eyes that we were being subjected to this kind of put-down. Then someone from our group said, "Thank you for agreeing to talk to us."

That statement was a signal to all of us that we should just leave. We headed out, and as we were nearing the dorm, Ibrahim, another member of the training group who was based in the United States, assumed a leadership role and summoned us to his room for a discussion. This was a safe place to vent and express the anger that was at a boiling point. After cooling off, we discussed how we were going to respond to what we felt was extremely insulting. We decided to call Ann and ask her to come to Brattleboro as soon as she could. We requested she also invite the head of training from the Peace Corps headquarters. We added that we were preparing to leave the program in addition to going on strike immediately.

We did not know how Ann would react, but we expected her to take it seriously since we mentioned a strike and the possibility of abandoning the program. Ann showed up in Brattleboro around ten in the morning the next day accompanied by the program planning director. We were sure that "all hell had broken loose." She must have alerted the supervisor that we had called Washington and were planning to leave the program midstream.

One thing I liked about Ann was that she understood the importance of interpersonal/cross-cultural understanding. The first thing she did was apologize for what had happened. Then she asked us to tell her the details and why we were so upset. Our designated leader proceeded to tell her the basic facts. Then, Al Amin, another Tanzanian recruited for the program who taught race and culture through literature at Rutgers University, seized the opportunity to represent us and explained why the supervisor's words were offensive to us. He added that it was not just an insult to us, personally, but also to Tanzania on the heels of trying to rebuild a broken relationship with the United States through the Peace Corps program. The discussion took about an hour. The program planning director apologized on behalf of her office and requested that we consider finishing the program as scheduled. She also announced that we were going to be moved to a hotel downtown and that she would compensate the time we had worked outside of the required eight hours a day. She conceded that staying in the dormitory didn't add value because we spent enough time with the participants during the day and we deserved personal time, as did the trainees.

What carried the day for us was the director's willingness to leave her work and travel to Brattleboro within 10 hours from when we called Ann. We also credited Ann for her sensitivity and understanding.

We agreed to go back to work immediately after the meeting. The Peace Corps trainees must have sensed there was something going on since we hadn't shown up to class and we told them to study and wait for further instruction from our supervisors. After lunch, we prepared to move to our new residence in town while our supervisor prepared to leave for a different assignment. Ann stayed with us for the remainder of the training period. This experience was culturally valuable. Despite all of this, I had plenty of memorable fun times in Vermont and the United States. My favorite story was a round trip from Brattleboro to San Francisco. I did not realize how adventurous this was, but it was an opportunity to experience America by traveling through it. There were two reasons why I wanted to make the trip. It started in Tanzania when I met a couple from the United States—Jean Brown and Charles Prael from Palo Alto, California. Charles and Jean were newlyweds on a world tour for their honeymoon. They knew my brother Ladislaus (also Ladi) who was a student at Stanford University doing graduate studies. Jean was an in-service grade school teacher at Stanford at the same time as Ladi, pursuing a master's degree in education. She and Ladi collaborated on several class group projects. Charles was an accomplished labor lawyer. His first wife had died a few years before his marriage to Jean, and he had four grown children, two boys and two girls. Through the group projects, Jean and Ladi became friends because Jean became interested in the education system in Tanzania. She wanted

Ladi to meet her husband. She invited him for dinner at their house. It was during dinner that Ladi learned about their plans for world tour. They mentioned that they were going to visit the Seychelles and Madagascar. Ladi encouraged them to visit Tanzania as well since they were going to be in East Africa. He volunteered my name as a contact person when they got to Tanzania. Both Jean and Charles were excited about the suggestion and added Tanzania to the list of places to visit. At that time, I was working at the University of Dar es Salaam as a researcher. When Ladi asked me to host the Praels, I immediately agreed. As a researcher, it was easy for me to reorganize my schedule to accommodate their visit.

Charles and Jean arrived in Tanzania in July and spent five days. They wanted to visit one national park, a primary school, and a teacher training college. The closest park for them, since they had only a limited time in Tanzania, was the Mikumi National Park, about three-hundred kilometers from the capital city of Dar es Salaam. I made the tour reservations, but I was not able to accompany them. However, I arranged for them to come to my house for lunch one Sunday afternoon. I was honored when they accepted because I knew this was their first visit to an African country and they might find our cuisine different from what they were used to. I took home economics in middle school, and now I was about to put my archived knowledge to the test. The menu was roasted chicken, boiled rice, boiled potatoes, and sauteed cabbage. For dessert, I prepared fruit salad with custard on the side.

The lunch was scheduled for 1 p.m. I organized a pickup from their hotel. I was delighted that they had accepted to come to my humble home, but I was also nervous about my menu. They arrived on time and after exchanging greetings we sat in my tiny living room for a drink. Later we moved to the dining table to eat lunch. The display on the table was impressive. They looked excited and kept saying thank you for the invitation. The glee on their faces assured me that all was well, and we settled down to eat. They took small portions, which was a safe way to go when you are not familiar with the food. But these were quite common dishes and, in my mind, international. When they went for seconds, I felt I had passed my cultural and international test. I was incredibly happy. They praised me for everything, including my creative dessert, which was an invention of mine. We made custard a lot in my middle school home economics class, which was very British in orientation. I thought fruit was easier to prepare than cake, and it also was an opportunity to display the variety of our tropical fruits. After lunch, we sat in the living room for a while, and I served coffee. Before they left, they invited me to their hotel for dinner, and we agreed the best time would be their last day in Tanzania. They were staying at the former Kilimanjaro Hotel in the city center, near the harbor with a spectacular view of the Indian Ocean. On the day after their visit, they left for the Mikumi Park tour. While Jean and Charles were in Mikumi, I visited the City Council and obtained a permit to take them to one primary school and the only teachers' college in the city. The council met all the logistics and provided an official to accompany us. My responsibility was to provide the introductions at the selected schools.

The trip to the park was successful; they saw a herd of lions and countless elephants. I joined them in the morning at the hotel to take them to the selected primary school where our official was to meet us. At the school, both Jean and Charles were taken by surprise at the preparations made for their visit. I had no idea what the visit was going to be like, but it turned out to be the couple's first major culture shock. The school had created a full itinerary for them beginning with a live school band and a welcome parade that started from the school gate. What was most surprising was the band starting to play as soon as our car pulled into the school grounds. Two students, a young girl and boy, were standing at the entrance, and they marched military-style to the beat of the drums behind the band and in front of the car. When the car stopped, the girl opened Jean's door and the boy, Charles'. The band stopped playing and the principal stepped forward. He shook hands with us and said welcome in both Swahili and English. He asked Jean to follow him while a staff member led Charles and me to the teachers who were standing near the students. The students stood in neat rows dressed in their uniforms of white shirts and blue skirts for the girls and blue shorts for the boys. The band resumed as Jean and the principal walked up and down the rows. This resembled the inspection of the troops except that the students were not holding guns. As soon as Jean approached a row, all the students stretched their hands out for Jean to inspect the cleanliness of their uniforms, shoes, and nails. This is a routine the students went through each morning before classes, an encouragement to learn and always maintain hygiene. At first, Jean was puzzled but then got the idea as she went through the rows. Back at her school in San Francisco, Jean could not stop talking about this experience and wondered how her students would have reacted to the process. When I visited her school on my first trip, this was the first thing her principal and teachers talked to me about. It was memorable to both Jean and Charles, but more so to Jean because she was a teacher.

The night before they left Tanzania, I joined Jean and Charles for dinner at their hotel. It was special because they selected the rooftop as the spot for our table. It was an open-air space where one could see the Indian Ocean, ships coming in, fishing boats going out, and the sheer beauty of the blue sky over the harbor. It was my first time in this hotel. The six-course menu was special, and the dishes were brought out one at a time. I must have used every imaginable utensil, a specific one for each dish. I had never had cold shrimp, lobster, let alone a cocktail before dinner and an aperitif after dinner with coffee. I was experiencing America in my own backyard through these newfound friends.

Jean and Charles were flying out the next evening, and I planned to go to the airport with them to see them off. They insisted it was not necessary, but I was adamant; it was our culture. They relented and I waited in the lobby for them. When they came down from their room, they were happy to see me, I was already a close friend. At the airport, we exchanged our good-byes. I casually said, "See you in America." They asked if I was planning to go, and all I could say was, "You never know." Little did I know it was a premonition. Two months later, in August, I was on the plane to the United States. So, this was why I was so eager to see where they lived. Jean and Ladi were also graduating that year, and I was excited to join them on their special day. Jean and Charles offered to let me stay at their home in Palo Alto, a five-minute drive to Stanford University.

The trip to California from Vermont started on Thursday morning. I had requested permission from Ann for time off to make the trip to California. She was gracious and allowed me to do that. The bus left from downtown Brattleboro, at six-thirty in the morning. I tried to observe everything we passed on the way. Our first stop was at lunchtime when the driver announced a bathroom stop and the opportunity to grab snacks at the nearby coffee shop. I decided to start with the bathroom. This was also my first time using public facilities at a bus stop in the United States. The facilities were genuinely nice, clean, and spacious. When I tried to open one of the doors to the stalls, it seemed locked. I tried three more with the same outcome. As I was wondering what to do, a woman came in and put some money through a slot on the door. The door opened, and she went in. What a shock that you had to pay to use a toilet. This seemed uniquely American. I read what it said on the door: "10 cents." I did not have 10 cents, so I waited for the woman to come out. When she did, I politely requested to go in after her since I did not have the 10 cents needed. She was kind and replied, "Be my guest." She held the door for me, and I went in. I had not paid attention to the culture of opening and closing doors in America. I was about to experience another culture shock. I could not figure out how to open the door to get out of the stall. I thought I needed to pull the door in instead of pushing it out. The more I pulled the more it became apparent that I was trapped in the stall. I remembered what the woman had said to me. It was the first time I had heard the phrase "be my guest." I thought it meant I was a guest in the stall until the next person with 10 cents came to get me out. There was space below and above the door. I thought that one of them might serve as an escape route. When I assessed both possibilities, it became clear to me that it was impossible. I was not a tiny kid with flexible limbs. I heard the driver announce our departure. I started to panic. I did not hear any footsteps coming in. Then I heard the second call for departure. Now I was more frightened because I was not a very sophisticated traveler in a foreign country. Stupidly, I also had left my purse on the bus. This meant I had no identification or money. I decided I was going to give the door a good push and then pull as hard as I could. It turned out to be a good move because the door flung open. Ha! I was supposed to push rather than pull and here I had spent almost ten minutes trapped in a bathroom stall. I quickly exited and washed my hands. Near the sink, was a big cylindrical object with writing on it: "dry hands here." As I was trying to figure out how this was done, a woman walked in and saw that I was puzzled. She said, "Push the button." I pushed it, and a loud noise started. Thinking I must have broken it, I left without drying my hands, looking back just in case the woman saw me and reported I had broken it. I was the last person to get on the bus, and I was grateful they had waited for me. I did not have time to grab anything to eat but I was happy to be on the bus. A woman sitting next to me was eating a sandwich and had a bag of chips. She said she did not care for the chips and asked if I would like them. I was not shy. I thanked her and accepted her offer. This was to be my only snack until I reached California.

The trip was smooth from there on. I marveled at the landscape and was excited that I was finally visiting the land of plenty. I recited my favorite phrase, "In California, we eat what we can and can what we can't." I was about to discover the truth of that. The bus pulled into the station, and I got off and saw Jean and Charles waving excitedly. I ran to them and we hugged, laughing and celebrating what had seemed impossible two months earlier. Everything seemed so big compared to Brattleboro. This was a city with skyscrapers, lots of cars on the highway—everything was extraordinary. I was finally in California. I remembered a song that was our favorite in middle school Sunday afternoon social hour, "I Left My Heart in San Francisco" by Tony Bennett. The nuns had a phonograph that they pulled out and played records for us while we danced all afternoon. We learned the waltz, rock and roll, slow dancing, the twist—you name it, the sisters taught us. We modified some of these dances with our African moves.

As we drove to Palo Alto, Jean pointed out landmarks, giving a brief history and their importance. Then Charles asked me how the trip from Vermont was. I said it was long, but interesting. I proceeded to tell them about being trapped in the bathroom. With the dramatization of it, both laughed to tears. I was laughing too, and as I laughed, I made them laugh even more. The same happened when I returned to Brattleboro and retold the story to my colleagues. It was hilarious and a major cultural experience. My stay in California was brief but memorable. I promised my host I would return. In truth, I did not know if it would be possible considering the cost. But as this story shows, no one ever knows what tomorrow holds.

Back in Brattleboro, I joined my colleagues to complete the training program. After the change in management, everything went smoothly. During the last week at SIT, we prepared the trainees for their program completion ceremony, which included skits and songs in Swahili. In the evening we had a big closing party. The trainees received certificates of completion and awards. The next day, we received our final pay and later that evening we left SIT. My colleague and I headed to the airport to return to Tanzania while our US colleagues took buses or drove to their hometowns.

Once in Tanzania, I returned to my research job at the University of Dar es Salaam. The trip was life changing, I started to see Tanzania in a new light. I saw how far we needed to go to be like what I had seen in America during my mere eight weeks. I wondered if Tanzania could ever get to such an envious position where the infrastructure was spectacular and people had access to basic life needs like good health services, education, agriculture, and many other things. I thought about my parents in the village who were still using a hand hoe to till the land, grow the crops, harvest, and carry it on their heads back to the village for storage. I thought of John Steinbeck again, *"In California, we eat what we can and can what we can't."* I thought of the villages where the uneven dirt roads were impassable, especially during the rainy season, a place where people walked miles to get anywhere. I recalled words from the then-president of Tanzania, Julius Kambarage Nyerere, who declared that his country was far behind the developed world and noted that while others walked, we ran to try to catch up, a world in which the American and Russians were trying to get to the moon while we were trying to get to the farm. It was a sobering thought, but no immediate solution was on the horizon. At that time, Tanzania was at war with Idi Amin, the president of Uganda who tried to annex a part of Tanzania in the western region. Amin had killed a lot of his people and now he was trying to mess with Tanzania. The Tanzania Parliament resolved that the nation should defend itself, and the president decided to send the army to the border to push Amin's troops out of the Tanzania territory. The effects of the fighting on the country were beginning to show in food and basic supply shortages because the country's resources were diverted to meet the needs of the war. No help was coming from the West. In fact, America sent help to Uganda through Libya and Kenya with the hope of defending Uganda. Amin came to power in Uganda because of a 1971 coup against President Milton Obote, who now was in exile in Tanzania. The explanation was that Obote had put his country on the same path as Tanzania, which was a socialist country in alliance with Russia and China. The Cold War was not being fought inside Tanzania.

While I was contemplating all this, something else non-political was happening in England at York University two months after my return from America. The sudden turn of events changed my life's trajectory forever.

Grandma Elizabeth (aka Malya)

Visiting home after moving to the USA: seated (middle) with my parent on each side.

With nine of my siblings attending Mother's funeral in 1995. Front row from left: Matilda, me, Father, Clemence, Wenceslaus. Back row from left: Venance, Anthony, Joseph, Cassian, Florentinit, Adelin.

My brother Ladislaus (left, with dad).

My nieces and nephews at Mother's funeral, November 1995.

Catherine, my middle school teacher.

PART 2

Opening the Doors to Success

What is life?
Life is a journey with start and end points.
Life is a partner you cannot run from. You end together.
Life is what you want to make of it. Benefit from it.
Life is all the beauty around you. Enjoy it.
Life is like a dream. Chase it, and when you catch it, recognize it.
Life is a challenge you cannot run away.
Life is a duty. Respect it, do it well and completely.
Life is a game you can win or lose. Plan to win.
Life is all that you promise yourself. Fulfill your vows.
Life is happiness and sorrow. Live it and overcome it.
Life is full of lucky spots. Recognize and be thankful.
Life is an adventure. Be brave and go with the flow.
Life is precious. Do not waste it; live it to the fullest.
I am grateful for how far I have come in my life.

— 9 —

Game Changer

One Monday afternoon, sitting at my office at the University of Dar es Salaam where I was assigned research work, I was trying to wrap up my day's work before going home when I heard a knock on the door. I was mentally exhausted as I tried to put the day's events into perspective. This Monday was challenging because of some personal problems that started Sunday, causing me to be late for work. More specifically, instead of being at my office at eight o'clock, I arrived at ten only to realize that I had left my office keys at home. To get into my office, I had to walk across campus to the housing department to request a duplicate key. I was already tired because I had missed my connecting bus to work and had to walk half a mile from the last bus stop to my office. I got the duplicate key and walked back, physically exhausted. I closed the door, spread the cushions from my chair on the floor and lay down for about a half-hour after drinking a couple of glasses of water. After resting, I straightened myself up, opened the door, and started working. I skipped lunch because I wanted to do some work before I headed back home, another hour on three different buses, from campus to the highway, another bus to town, and finally a bus to the suburb where I lived.

The knock on the door came as I was finishing organizing some data my assistant had collected for me. It was the secretary. She had been sent by the director to let me know he wanted to speak to me. I took the stairs from my office on the second floor to the director's office and found him at his table, looking through some files. He offered me a seat and then said, "I have some good news for you."

"I could do with some good news. Today has not been my best day," I replied.

"What happened?" he asked.

"Nothing worth talking about, just some personal issues," I replied.

"Well, you need to start getting ready to travel to York University in England because you have been given an opportunity to start your Ph.D. program," he said.

I did not respond immediately because I was in such a state of shock. To begin with, I was not expecting to go for my doctorate this soon because I had just joined the University of Dar es Salaam's research department. The university had a professional development track for its staff, and it operated on a seniority basis. Several senior colleagues had not yet been assigned. Furthermore, based on the traditional schedule, I was to go to Australia in three years to study for my Ph.D. in linguistics. In addition, we had a staff member who was already at York University, and the exchange program with this institution was on a one-to-one ratio. Sensing my shock and surprise, the director continued talking. He indicated that I did not have much time to prepare because I was already two weeks late. What is more, he knew it was easier to select me to go because I had just come back from overseas, already had a passport, and would not need government and health clearances, like vaccinations. Unlike traveling to America, I did not need an entry visa because Tanzania was part of the Commonwealth countries. I asked whether York University had changed its one-to-one exchange policy, reminding the director that we had another colleague already there. He paused for a second and then explained that this colleague had decided to return home without giving much warning to York University. This created a problem for the department because part of his scholarship included teaching Swahili at the university. The director thought that since I had just been involved in a teaching program in the United States, it was easier to send me because I would be able to hit the ground running. At this point, I realized I had no choice but to accept the offer and start thinking about my trip to England. York University had sent instructions, including where to get the plane ticket and what to do when I arrived in London to catch the train to York. Once I arrived at York, there would be someone at the railway station to drive me to the school and get me settled.

I returned to my office where I sat at my desk for ten minutes going over what had just happened. I remained puzzled at this turn of events. I knew I had a week to get ready and leave for England. It was easier to prepare for this trip because of my earlier US experience. However, I was more apprehensive because this time I was traveling alone, and I had no clue about what had happened for my colleague to leave England in such a hurry. Nevertheless, I was optimistic and somewhat excited about the opportunity to get an early start on my Ph.D. studies. The week went by quickly, and I soon left for my new adventure. When I left Tanzania for England, I was not sure how long I was going to be gone. I also was not sure how well I would do as a student at a foreign institution. The only thing for certain was that I was being rushed to York University in England to replace my predecessor who had abruptly decided to shorten his stay and return to Tanzania.

At Heathrow Airport, I followed the instructions sent to me and headed to the train station where I had a reserved ticket. The trip was pleasant; the countryside was lush with scattered shrubs and sheep grazing. Having grown up in rural Tanzania, I appreciated the landscape and its calming effect. The rolling hills reminded me of the times I went to cut grass for the cows or looked for firewood. But this was not Tanzania. It was a foreign country, the very country that colonized Tanzania before it attained its independence in 1961. Throughout my stay in England, I kept comparing what I saw to my home country and marveled at the differences.

At the York train station, I saw a young man holding a sign with my name on it. The school had sent him to pick me up. I was grateful. He helped me with my luggage and dropped me off at my assigned residential hall. He must have been instructed on what to do because he was like a hotel attendant, showing me what I needed to know and making sure I was well settled in. Once satisfied that I had everything I needed, he left. The only thing the young man did not cover was how to find food. There was a small kitchen in the corridor of what looked like an island of three rooms; mine was one of them. I learned later that on each floor, one of the islands was designated to the graduate students. The kitchen had everything one would need to make a meal. Despite the furnished kitchen, I still did not have food to cook. Later that evening, I heard a knock on my door. Two people introduced themselves as Sylvia and

Mark. Sylvia was from Jamaica and Mark was from Belgium. Sylvia was doing a Ph.D. in chemistry and Mark in physics. They told me they had been expecting me and were happy to have me as their neighbor. They had established a kitchen-sharing system where they cooked and ate together. They invited me to dinner, assured me of breakfast in the morning, and invited me to use whatever I needed from the kitchen until I was able to stock my own locker. This was a wonderful way of sharing cultures through food and fellowship. I was surprised by their generosity and excited that I had found an instant, friendly family. After dinner, we sat in the kitchen and talked for an hour about my country and theirs. For the rest of the academic year, until they graduated, I was assured a family environment of love and friendship. We shared each of our culture's dishes, and I learned a lot about Belgian and Jamaican cuisine. I enjoyed rum and many fine wines. I considered this experience an exceptionally soft landing in England, one I was grateful for.

There was a welcome package in my room with an introductory letter and information about the Languages and Linguistics Department. I also was notified that someone was arriving in the morning to take me to the department. Around 8 a.m., I heard a knock on the door and when I opened it, a tall, elegant woman greeted me and introduced herself as Joan. She was my supervisor and had come to take me to the department. I had already eaten breakfast and was ready to go. I was introduced to the department head and then was shown where my first linguistics class would be later that day. Joan took me to my Swahili class, which she was teaching until my arrival. She introduced me to the class and allowed me to sit in and observe. I took over the teaching the following day.

Because I intended to earn a doctorate, Joan made an appointment the next day to discuss my academic status. I gave her a copy of my certificates: undergraduate, teacher training, and master's in linguistics, all received in Tanzania. What followed took me by surprise. She looked at me. I had written my master's thesis in my native language, Swahili, and she said this was not an acceptable language at the academy. I was informed that I had to obtain another degree, a Master of Philosophy. Although Joan's first language was British English, I knew she spoke fluent Swahili, she researched for her Ph.D. using Swahili, and she could read my thesis and make an evaluation of its quality to allow me to proceed with the doctorate program. Furthermore, the sole purpose of my scholarship was to teach graduate and undergraduates Swahili. This was a big disappointment, but I had to accept her decision although I was proud of my work and the fact that I was the first person to ever write a master's thesis in a language other than English. But here, this accomplishment was a detriment rather than an advantage. I took the setback positively and embraced the idea of a second master's degree with a different title. To me, it was yet another feather in my hat.

I enjoyed teaching Swahili to graduate and undergraduate students at York while I took new courses required for the master's degree. Tuition and accommodations during my studies were covered by a British Council scholarship through the university. This was established as part of the exchange agreement to enhance education in former British colonies. The scholarship included a small stipend, paid at the end of each term, not monthly, which was approximately 400 English pounds (about $540 per semester). The stipend's schedule was a major adjustment for me. I had to learn how to budget whatever I had to last an entire term (ten weeks). I set my budget to thirty-five pounds a month (about $74) for food no matter what. My eyes were always on sale items, a habit I shamelessly keep to this day.

The thirty-five-pound-a-month limit gave me a maximum of 105 pounds (about $142) of food money for the entire school term. I was able to save 295 pounds (about $215) each term for a total of 885 English pounds (about $1,196) in three terms. I planned to go to Tanzania in the summer to visit my family. I made sure I saved enough money to allow me to survive the ten weeks of the term before the first stipend was distributed. Because I was sharing meals with the other graduate students, it was much easier knowing I had two days of cooking and one day on my own when I could have something simple. Bread and stew (meat, chicken, or fish) were my favorite meals. I would vary the stew recipe each day to make it different and interesting.

During this time as a doctoral student, I stayed in touch with my California friends. Whenever they wrote me, they never failed to ask how I was doing. I was candid about my budget restrictions. While I thought I was simply telling stories about life in England, my friends looked at this stipend as a major hardship. In their following letter, they sent me a money order for two hundred and fifty dollars. This was a great deal of money. I was embarrassed but at the same time very thankful. Interestingly, they made it a point of sending me support every two months. This was exceptionally generous, and I never forgot their friendship.

After my first year in England, I thought I had mastered the British system at York University. But a couple of surprises awaited me in my second year. In my mind, this was my last year in the master's program, and I planned to advance to the Ph.D. program. Such a step would not require me to defend my thesis. I just had to complete all the coursework, research, and write my thesis. Because I did not own a typewriter or know how to type, I wrote everything long-hand. I found a typist who doubled as my editor. My challenges in English were minimal but I did not want to take anything for granted since it was my third language, one I started to learn in the fifth grade.

By the beginning of the fall term of my first year, I had sorted out my teaching schedule and planned out my research and writing. Each day I would have all my teaching obligations completed by lunchtime, thus leaving plenty of time in the afternoon and evening for other activities, like research and writing. This schedule allowed me to travel to London on select weekends to visit with a family from Guyana that had adopted me. I will not do justice to my story if I do not talk a little bit about this family.

It was a sunny Wednesday afternoon when I met this stranger, a petite woman named Vera. She was confirming her return ticket to England at an airline office in Dar es Salaam. I was at the same office to purchase a ticket to go to Nairobi. Her light skin tone did not reveal her race since we have many mixed-race people in Tanzania from different parts of the world. She did not look like someone who had been in Tanzania for long, but she seemed confident and secure. Our eyes met several times and she smiled at me. I moved from my seat to sit next to her, and we exchanged greetings. That is when I learned she was from Guyana but now lived and worked in England. She had come to Tanzania for a few days but did not specify who she had come to visit or whether she was here just enjoying the sights by herself. We ended talking for an hour before we parted. We exchanged addresses and promised to write to each other when we had a chance. She extended her arms, and while hugging, I said, "See you in England."

"When are you coming?" she asked.

"I do not know. The chance to travel to England is always there," I replied.

Honestly, I had no idea why I said that. Indeed, it happened like magic. But I have always believed that everything happens for a reason. It was a big surprise to Vera and me when I ended up in England four months later. She had a sister and two daughters in England and several nieces and nephews whom I got to meet during Christmas and other holidays. The opportunity to study at York University was what reunited us.

I arrived in England on October 8 and waited for a month before I notified Vera. I sent her a letter using the address she had shared with me. There were no cell phones then and no phones in the residence halls. Four days later I received a reply and an invitation to visit when I had a chance. That is how I became a member of Vera's family. She is now retired and spends part of her time in Guyana and part in England. I am still in touch with her and her family and in 2018, I visited Guyana, her ancestral home, and met many of her cousins, brothers, sisters, nieces, and nephews. Vera is now ninety-five years old, but she looks like she is only sixty. I cherish her because she gave me a family in England during my studies, allowing me to stay focused, and not feel so homesick. It would have been hard to tolerate the weekends I had to spend alone in the residence hall when most students went home, leaving on Fridays and returning on Sundays. It was Vera's friendship and nurturing that helped me get through the two years at York University.

Many of the graduate students were more adventurous than I was and took trips around the country. I did not have a TV, but I bought a small radio. I listened to classical music and BBC News. This little radio led me to buy a phonograph and a lot of vinyl records and cassettes of different artists including Mozart, Beethoven, and Bach. These music formats have become relics, but I still have many vinyl records and audiotapes that I have kept all these years. One of the graduate students had a TV, and occasionally I would go to her room to watch *Panorama*, a BBC program that was on at 5 p.m. every Sunday. The show displayed different churches around England with religious choirs in the background. The program fascinated me and became something to look forward to every Sunday. Another favorite program was the American sitcom, *Dallas*. This dramatic family saga came on from 7 to 8 p.m. Saturday. I loved it so much that I was willing to forgo any invitation that would force me to miss an episode. When my friend graduated, she gifted me her television, which allowed me to enjoy the weekend in my room free of boredom.

I cherished anything that sparked my interest and kept me from thinking about the numerous challenges and difficulties I encountered at York University. It seemed like there was no shortage of surprises that crept up throughout my stay in England. I had told myself that I needed to find a way to stay focused and realize my goals, especially after my supervisor's refusal to accept my thesis because it was written in Swahili. The next bombshell came in the second term of my second year at York and had to do with my graduate teaching fellowship.

It was late one afternoon in October 1980 when I was summoned to the department head's office. First, he asked me how I was doing and if I knew there had been some changes in my studies. I said no. He then broke the bad news to me. My scholarship would end at the end of the academic year because the British Council would no longer make it available to foreign students. He added that I had two choices, to finish at the Master of Philosophy level and return to Tanzania or pay my own way for the doctorate program. Neither choice was good, and I was at a loss as to what I was going to do. I came to England to earn my Ph.D., not a second master's degree. At this rate, my experience was a waste, and I should have waited for my turn to go to Australia.

While I thought this was all I had to worry about, I was confronted with another surprise two weeks later, a week before the term was over. The business manager informed me that my stipend would be cut by half because the department did not have enough funds for all the teaching assistants. I was shocked and disappointed because at the beginning of that term, the department had hired a new teaching assistant to start teaching Chinese. Surely, they must have completed this hire with full knowledge of the budget. I was already on a shoestring budget and now it was going to be slashed in half. Later that day, I asked one of my colleagues who taught Creole if she had heard the news about stipends being cut by half. She was not aware of it, and she went to see the department secretary to inquire about the new budget. She was surprised to hear that she was not going to be affected because she was in the Ph.D. program already. The secretary clarified that only one or two students would be affected. When my friend brought back the news, I could not hide my sadness nor avoid thinking that I had been singled out and perhaps no one else was going to be affected.

The next day, I went to the office and asked to see the department head. The secretary was kind enough to let me see him immediately. This was a tough visit because I had never been comfortable in his presence. This discomfort started the very first Monday after I arrived at York. He taught a graduate-level class on sociolinguistics, Language Variation. I had missed two weeks of classes, so I was quite behind in what was going on in the class. The topic for the day was phonetic transcriptions. He called me to the front of the class to transcribe, on the board, a statement he had made. I provided the best answer I could from what little of the lecture I had heard and what I knew from my linguistics background. He then asked me why I thought that was the correct answer. I could only respond that in phonetics one uses the IPA (international phonetic alphabet) to represent the approximation of what was said. His question sent a chill down my spine. I told myself not to dwell on it, that he did not mean any harm. However, since it was my first day in class, I could not help but take his question personally and feel intimidated. So, I entered his office already scarred by these negative feelings I had harbored for over a year but was determined to get to the bottom of this sudden stipend reduction in the middle of the term. As soon as I entered, he said, "Yes, come in, take a seat."

I said hello and then proceeded to ask my question: "I was notified yesterday that my quarterly stipend will be cut in half. I was wondering if I could get some clarification from you."

"Oh, it is accurate," he said. "You are now sharing your term stipend with the new teaching assistant for Chinese. You know Chinese is an important language now."

"But this is a significant cut," I responded.

"I know, but that is how it is going to be," he added.

He took his eyes off me and went back to what he was doing before I entered. Sensing he would continue to ignore my presence, I left, disappointed and with no recourse. The secretary looked at me as I exited. I turned around to thank her for fitting me in. She must have wondered how the meeting went but she did not look like she was going to ask. Back in my dorm room, I sat on my bed for what could have been at least a half-hour. I started to plan my next steps and my priorities to ensure that I finished my MPhil and would then move on.

Shortly after this incident, my brother Ladi visited me on his way from the Philippines to Nairobi where he worked. I told him about my predicament, and he casually suggested that I look for schools in the United States. He was familiar with graduate work in the US after completing his master's in Education and Communication at Stanford University. The idea was inviting. I had been in the US in 1978, visited Stanford to attend his graduation, and connected with my friends Jean and Charles in Palo Alto, who were now supporting me both emotionally and financially.

I took his suggestion to heart but did not do anything about it until February 1981. I had approximately four months to finish my second master's degree. My thesis was being typed and edited at this time with a plan to complete and defend it in late June. As such, I had time to do other work. I decided to go to the library and research schools in the United States. I got information about UCLA and

Stanford and sent a handwritten letter to each. Stanford was the first to respond in about a week. They encouraged me to apply for the 1982-83 academic year because they had closed their admission process. Shortly after, I received a postcard from UCLA thanking me for the inquiry and nothing else. However, the postcard had valuable information: courses taught, assistantship possibilities, and other work-study fellowships. I wrote back and indicated that I was teaching Swahili at York University and that if there was a need for a teaching assistant, I would be happy to oblige. I did not hear anything for a couple of weeks. One day after a meeting with members of my thesis committee, one of them pulled me aside and asked me if I had applied for graduate school at UCLA. I said yes, and he proceeded to tell me that one of his friends, a professor of phonetics and a graduate coordinator at UCLA had inquired about my ability to do graduate work. They had contacted the department first and were referred to my major professor who told them my English was poor. The UCLA professor, also British, decided to seek another opinion and contacted his friend who I had listed in my application letter as a member of my thesis committee. I also had taken two terms of coursework with him; he taught psycholinguistics. He told me he was surprised that someone had told his colleague at UCLA that my English was poor. He assured me that he had encouraged them to admit me as a new graduate student and that my English was as good as any of his other students in the classes, he taught. He lifted my spirits, and I started getting hopeful. "Perhaps," I thought, "there is a light at the end of the tunnel." A few days later, I received a letter from UCLA. I was nervous and did not open it until after dinner. I was bracing myself for another disappointment. I could not contain my joy when I read the first sentence, which said, "Congratulations, you have been admitted to UCLA to begin classes in the fall, 1981."

I did not bother to read the rest of the letter until morning. I jumped up and down with joy, cried a little, and said my "thank you" prayer like a good Catholic. To go to sleep faster, I usually drink a cup of coffee, contrary to what people say about caffeine. I made a cup of coffee, and ten minutes later I drank enough to fall asleep. All night long I had dreams about California based on my short experience visiting Stanford.

When I woke up, I remembered some of the dreams, and I was confused as to whether they were real or not. I saw the letter on my desk, opened it, and read the first sentence again. I read the rest of the letter and tried to imagine what the next phase would be like. I was excited but determined to keep this close to my chest. I was afraid someone might undermine this incredible opportunity. I washed up and went to morning Mass at the school chapel. The priest knew me well because I frequently attended his sermons. I also had told him about my dilemma and asked him to pray for me to find a good solution. After Mass, which was for me one of thanksgiving, the priest asked me how things were going. Since I had asked him to pray for me, I felt obligated to let him know that his prayers had been answered. I felt that I could trust him with my secret.

"Things are looking up, Father," I replied.

His face beamed. I told him the details, and he smiled even more.

"I knew it was going to end well. Congratulations, and keep up the good work," he said. The priest was American and allowed hugs. I hugged him and said, "Thank you, Father."

I returned to my dorm room, had a quick breakfast, and then left for class. I felt the urge to share the news with my students, but I restrained myself. From that day, I started planning for my departure to the United States. I had renewed enthusiasm to finish writing the thesis and do the necessary preparations for my departure.

The lucky stars were still on my side. A week later, I received a package from UCLA. Inside the package was a letter outlining all I needed to do before my arrival at UCLA. They said the embassy in London had been informed of my scholarship and all I needed to do was visit the embassy on the date shown in the letter to get a visa stamp on my passport. I did not have to fill out any paperwork. Also, while in London, I should go to an address included in the letter, the American Airlines office, to pick up my ticket. As if this were not enough, they told me to find a moving company that could ship anything I wanted to bring with me. UCLA was going to pay for it. I still wonder why I was so lucky. I do not think this was common practice, and I have never heard of it from any other student, foreign or domestic.

I had new energy. All I could think of was finishing my term, defending my thesis, and going to California. As it had before, John Steinbeck's book *The Grapes of Wrath* echoed in my mind: "*In California, we eat what we can and can what we can't.*" Well, my wishes were about to be realized. I was about to start a new beginning in California with a clear, achievable goal: to get a Ph.D. in linguistics.

The term ended well and without interruption; I successfully completed my thesis. I had to defend my work before four committee members and one external examiner from SOAS (School of Oriental and African Studies). I was nervous that day because earlier when I had run into the department head, he warned me about the external examiner.

"Good luck on your defense this afternoon," he said. "The external examiner is very tough, and she failed your major professor on her Ph.D. defense. It will be interesting to see if she passes you," he said.

I knew my major professor was writing her Ph.D., but I did not know that this was a rewrite after a failure. So, I went into the defense that afternoon armed to fight, to pass. The examiner arrived late; she had come on the train from London. To my benefit, she was also feeling ill with a bad cold, but she was determined to complete what she had come for. The defense started with me summarizing my work and answering questions from the external examiner. In comparison, the committee structure at York University was different from that I found in the United States. In the US, the committee assumed the role of an examiner, while at York University, the committee had a supportive role that allowed them to ask alternative questions when the candidate appeared to not understand the external examiner. During the defense, I was extra nervous and tried hard to impress the external examiner. Halfway through the Q&A, the examiner looked me straight in the eye as if she were ready to ask me yet another tough question. I was surprised when she said, "Be calm, you are doing fine. I really like what you are talking about, and I would like to learn more because I am doing research in this area for a conference paper."

I let out a big sigh of relief. After this comment, the Q&A turned from an interrogation to a forty-five-minute conversation. She then sent me outside while the committee deliberated. Shortly after, my supervisor came out to get me. The examiner stood up, shook my hand, and said, "Congratulations, you have earned your Ph.D."

While I was baffled by her statement since I knew it was MPhil, my supervisor jumped in and said, "Oh no, it is just a Master of Philosophy degree (MPhil)."

"That's fine, the work she has put into this thesis deserves that title," the examiner responded.

I thanked her and shook her hand as hard as I could. If it were America, I would have jumped up and hugged her. I continued to shake hands with each of the committee members, thanking them for their support. Then another surprise, since it had never happened to any student before, the examiner asked my committee to join her for dinner in town to celebrate my success. She indicated that she wanted to hear more about my work and about Tanzania where she had spent some time as a graduate student doing research for her doctorate. I had no words for this surprising gesture, but I was happy to accept her kind invitation. At dinner, my psycholinguistics professor, known by the students as Dr. G., asked me, "What is next for you?"

I suspected his friend from UCLA had told him about my acceptance into their Ph.D. program, but I did not want to be presumptuous and assume.

"I leave for America soon to start Ph.D. work at UCLA," I responded.

It was a big surprise to my supervisor. Everyone at the table congratulated me and wished me success at UCLA. My supervisor asked, "How long have you known this? You did not say anything about it. Congratulations!"

"I am sorry, I was busy getting the thesis ready and kept everything else at the back of my mind," I replied.

"Good for you. You will enjoy California," she added.

Dinner was superb. Of course, this was a high-end meal for me, a graduate student on a strict budget. The thought that my external examiner decided to celebrate my final days at York with a lovely dinner was still overwhelming. After dinner, Dr. G. offered to take her to the railway station and asked me if I would like to ride back with him so he could drop me off at school. This gave me an additional opportunity to continue the conversation with the external examiner and to see her off. At the railway station, I thanked her again for her time and for the lovely dinner.

The master's thesis defense became the best way to both end my stay in England and begin a new chapter on another continent. The defense was on Thursday, and I planned to leave on Monday, which left me with plenty of time before my departure. There were enough days to celebrate my graduation and to break the news to my friends and students that I was leaving the United Kingdom for America. My students were sad because I was not going to be their teacher the following term, but happy for me to pursue my dreams. I already had my visa and ticket and most of the things that I wanted to be shipped were packed in trunks. I needed to find a shipping company to pick up the trunks, for I planned to take only one suitcase and a small bag as my carry-on luggage on the plane.

Before the last day, I visited with my supervisor, the department head, and the staff to thank them for their help during my two years in the department. The news that I passed the defense and that I was leaving for America to complete my Ph.D. was spreading like wildfire. The secretary was the first person to greet me in the office and she stood to congratulate me and shake my hand. The department head must have heard her because he came out of his office and walked toward me saying, "I heard the news. Not only did you pass with flying colors, but you are also going to an exceptionally good school in America. How did you get that lucky?"

"I just applied, and they accepted me," I said.

"Congratulations, and good luck in America."

"Thank you. It has been nice to be at York University, thank you for your help and support," I said.

I made more rounds in the department before completing my visit with Dr. G. He was still excited about my transition to America and invited me to join his family for dinner on Saturday. I graciously accepted and, that weekend before I left, I met his wife and two young sons. I thanked him for his recommendation to UCLA and promised that I would make him proud. Looking back, I kept my promise. On Friday, I made a trip to London to visit with my friends who had taken me in and cared for me for two years. They were happy to see me and to know that I had passed my defense. They were also happy for my upcoming adventures in America. Vera had lived in New York, so she knew quite a bit about the United States.

I had finished my teaching responsibilities for the term and prepared to leave York for California. When the day arrived, I took the train to London's Heathrow Airport. I left early in the morning to make sure I had enough time to navigate the airport without the risk of missing my flight. All went well and I finally left England for America.

— 10 —
THE SECOND TRIP TO AMERICA

Like the other trip, my second journey to America was a fluke because my original intention was to get my Ph.D. in England. I looked back on how far I had come, and it seemed like a road trip to nowhere with a great deal of anticipation for what this new chapter had in store for me. I knew what the starting point was but had no clue what the endpoint would be. Once in America, I used to watch the fantasy drama *Highway to Heaven*, featuring Michael Landon, Victor French, James Troesh, and Margie Impert. I pictured myself on this highway as it stretched for miles with no end in sight, but I preferred to call it the endless road or the road to the unknown. I always tell my friends that I believe with my whole heart that everything happens in our lives for a reason, even accidents or unpleasant events. They contribute to our thinking process and how we act or react to pivotal moments in our development. Indeed, coming to America for a second time was not an accident, it was an unplanned yet purposeful event.

It was 1981, three years after I arrived in England. Because this journey was from England to the University of California, Los Angeles, I considered my stay in England a brief stop en route to my destination. America was planned for me, and all this time I had been following a guiding star to where I would be for the rest of my natural life. Looking back, California was also another brief stop to help me gather the energy and stamina to get to my final post, Georgia. Every time I decided to leave, I always returned because I could never get Georgia off my mind.

But before Georgia, the California stop was full of drama and adventures that were both rewarding and challenging. At times I thought I was not going to make it in America, but the sun kept sending rays of warmth that allowed me to continue the journey.

I arrived at Los Angeles International Airport from England around five-thirty in the evening and was met by my supervisor, Professor Thomas Hinnebusch. He took me straight to his house for dinner and to meet his wife, Claudia, and their children. After dinner, he and his wife drove me and my luggage to campus in their family van. Tom had all the information I needed to settle in my dorm room. Before leaving, his wife offered to meet me the next day to take me sightseeing. We drove down Sunset Boulevard, around Hollywood, and through Westwood, the town where UCLA is located. Finally, we ended up in Santa Monica where we had lunch at a small restaurant. Santa Monica was active with many people shopping and dining. Claudia also showed me a bit of Venice before taking me back to campus.

The time I spent in England became an asset for my second coming to America because it prepared me and reduced the culture shock. I felt I could handle anything that was different even though California was quite different from Yorkshire. The experience in Yorkshire was more instructive than Brattleboro, Vermont, my first introduction to America. There was more diversity in California, a true melting pot that we hear so much about in diversity discussions. There were people from all corners of the world compared to Brattleboro, where most of the people were Caucasian. I learned later that there were places in California where towns were named after a specific ethnic group, like Koreatown, Chinatown, and so forth. East Los Angeles and East Palo Alto, two places I spent most of my time in California, were populated by Black Americans and Hispanics. Most African students at Stanford who were not in university housing could be found in East Palo Alto. I got the impression that if you were Black and looking for a place to live, you went to the eastern part of a city. But I was proven wrong when I went to Philadelphia and found that the community on the east side of the town was not Black but mostly white. The Black community was concentrated on the west side of the city.

Once settled at UCLA, I explored different communities around the school and discovered that, except for the Hollywood and Brentwood areas, the population was extremely diverse.

My first day on campus was awkward. UCLA is a big school that sits right in the middle of Westwood with many departments spread from the northern to the southern end of campus. I had not ventured into Westwood by myself, and I was expecting to find someone who would take me around. I was thinking like an African, making a cultural assumption that the locals would automatically take care of me since I was a visitor (or, as I felt, a stranger) and help me get familiarized.

With the size of the campus in mind, I had anticipated difficulty finding my home department (Linguistics), but I was surprised the next day when I easily found the building near Meyers, the graduate residence hall, on Hilgard Avenue. At the office, I met the department secretary, Ana, who gave me a brief orientation. Since I had a scholarship, she advised me to open a bank account as soon as possible. I asked her where and how to open an account and she proceeded to give me a street map and added that there were several banks I could choose from. I am not good at multiple-choice exercises. Not knowing the banking culture in the US, I was not sure how I was going to handle it. I had not met any other students because I had arrived before opening day and most American and/or continuing students were still on summer recess. It was not the age of cell phones yet, so I could not call my supervisor or his wife, who was now a great friend, to help me with this unfamiliar task. In any case, I was determined to conquer my fear of the unknown. Though I did not come from a culture of reading street maps, I took the map. I planned to get a general sense of direction and then explore while I established landmarks for future reference. Based on the drive with Claudia, I had a fairly good idea of the direction I was supposed to go. I started my walking tour of Westwood, exploring and familiarizing myself with the city.

    I strolled leisurely through campus exploring what was going to be my new home for a few years. I finally arrived in Westwood. I did not bother to check all the banks on the secretary's list. The first one I saw was Bank of America and I went in. I told myself I could not go wrong with this one because it said "America," so the government must own it and my money should be safe. That's how it was in Tanzania, where the government owned and ran everything.

The inside of the bank did not look anything like the banks at home or in England. I stood by the door bewildered. A nice elderly man saw me and came toward me. He said hello and added, "You look new here. Do you need help?" I was so glad and accepted the offer, though I was still unsure how to react.

"Yes, I just arrived here from England. I need to open an account," I replied.

"Okay. Just stand here and one of the nice ladies behind the counter will call you," he said.

"Thank you very much," I replied.

The man left and I stood where he had directed me. One of the clerks summoned me to the counter. "How can I help you?" she asked.

"I am a new student at UCLA, and I need to open an account," I replied.

She told me I needed a UCLA identification card. I did not have one, but I had my passport with me. She accepted that and proceeded to fill out some paperwork for me. She read the rules and requirements. I needed a minimum of $50 to open the account. Luckily, I had cash with me and gave her $100. She wanted to know if I wanted the change back, but I asked her to deposit all of it. After all the paperwork was done, she gave me a savings account booklet. I did not consider opening a checking account then because I thought only businesspeople had checking accounts because they had a lot of money. I also was keen on earning interest for my money, which was what I was accustomed to both in England and Tanzania.

Completing this exercise was like passing my first quiz. I felt immensely proud of myself. Instead of heading back to my dorm, I decided to explore the Westwood shops and food courts. I saw a place where the sign read "Hot Dogs." I was shocked and thought to myself, "These people eat dogs?" I decided that it was going to prove whether I was in a friendly environment. I came to learn that this was sausage in a bun. When I wrote home, I had to tell the story, which sounded funnier to Swahili speakers when translated. because the word is associated with fire or heat. For years, we made jokes about it and laughed to tears. Another funny food story was when I saw a sign for "Taco Bell." For someone hearing it and not realizing the spelling, this was hilarious. One of my friends in Tanzania said, laughing hysterically, "They hang bells on people's bottoms?" they asked.

"No stupid, this is a type of food. It is like *flatbread,* but it is crisp. They roll meat and other stuff in it before they eat it."

"Have you tried it? Have you eaten meat from someone's bottom?" one friend asked.

They were just having fun with me.

"No," I said. "Maybe one day when I have more courage," I replied.

I must admit that, to date, I do not care for tacos, which I have tried at least two or three times. It is not for the reasons my friends were laughing. I just have not developed a liking for them, but I do enjoy watching others eat and enjoy tacos.

After my first California excursions, the weekend brought a lot of new and returning students to campus. I got to meet several, and some were in my Linguistics Department. Interestingly, all of us were on some sort of scholarship, which was an asset for me because I could pick their brains on how everything worked, things like when we got paid, what was covered or not covered by the scholarship, and so forth. One would assume that such information would have come from the secretary. She thought I knew because I did not ask her, which is what is expected in American culture. Not asking is part of my cultural background where we revere authority and are fearful of being presumptuous.

Meeting fellow graduates was made easier by having a common room and dining room in the dormitory. I quickly bonded with some of the graduate students, and we always sat at the same table for our three daily meals. I met other international students at the Newman Center, which served the Catholics on campus. It was across from the dormitory in a convenient location on Hilgard Avenue. The center's members were very friendly and made sure new students were paired up with veteran students and staff. I liked this arrangement for settling into my new home and probably was the major reason I retained my membership at the center for the rest of my time at UCLA. Membership allowed me to participate in their summer programs for youth. One summer after my first year, I served as a counselor at a camp known as UNICAMP (Universiade Estadual de Campinas). It was my first experience sleeping in a tent for a week as well as purposely running around the camp, in the bush, with nine- to ten-year-old kids. It was a great summer experience away from home.

My dorm at UCLA had double-occupancy rooms. They were spacious enough to have two standard beds, two closet spaces and two small desks and chairs. Because I arrived early, I was in the room alone for a few days. When the rest of the students started arriving, I was eager to meet my roommate. No names were provided ahead of time. My roommate did not arrive until Sunday evening, the day before school officially started. The evening highlighted significant racial differences. I did not see it coming, but when it happened, I was left dumbfounded.

It was around six-thirty when a young, white, female student opened the door and walked in with her suitcase. She looked surprised to find me in the room. I said hello but she did not respond. She put her belongings on the bed, opened the closet on her side of the room, and stared straight in. She had long brown hair that fell below her shoulders. She gathered her hair up as if wanting to tie it into a bonnet, but instead held it high above her head for about five minutes staring straight at the closet rails because there was no mirror inside. At this point, I was not sure what to do or say. She turned around, threw a quick glance at me, and then picked up her suitcase, placed it in the closet, and locked it. Then she left the room. I did not see her again that evening, and I did not see her at the dining room during dinner. Late at night I found her sleeping in the bathtub in the girls' common washroom. I was very disturbed because I did not know what had upset her so much. On Monday morning, I also did not see her. By the time everyone went to the bathroom to wash up, she was gone. Later that afternoon, when I came back from class, I found her closet wide open and her suitcase gone. Finally, I got the message. I was sure she was gone and was not planning to come back. I now had a room to myself unless the school assigned me a new roommate. Interestingly, no action was taken, and I kept the room to myself for the entire year.

I must admit that, although I was not sure what to make of this encounter, I was happy I had a whole room to myself, complete with two beds, two closets, two desks. It was like a room in a hotel but better because I could enjoy double of everything. Despite this advantage, the experience overall left me concerned that race and ethnicity might be an issue at UCLA, a place where I was beginning to feel comfortable in less than a week. I did not talk about the experience with anyone until toward the end of the term (we were on the quarter system) when some classmates came to my room and discovered I had no roommate. They thought the college had favored me but there was no real reason for that. They thought this because my graduate status had been adjusted, unlike for other new foreign students, and I did not have to take some remedial and required courses. Also, they did not require a reading of my MPhil thesis to determine whether I would be admitted to the doctorate program. Waiving these requirements meant I could start attending classes that would contribute right away to my Ph.D. dissertation. To satisfy my classmates' curiosity, I told them about my encounter with my assigned roommate. Revealing her reaction when she met me for the first time and the fact that she chose to spend the night sleeping in the bathtub shocked them. One of my friends had seen her and wondered what she would do next. Of course, none of us saw her again. One of my friends told me I should not take it seriously or worry about it because she was one in a million of many good people in America, and I should just plan to enjoy my stay. That was encouraging and refreshing.

It was helpful to stay on campus the first year. After that, I had to look for accommodations off campus. I did not know much about apartment living, so I depended on students who lived off campus and had some experience in this. One friend found a South African student in the education department who was looking for a roommate to share an apartment. We met over lunch and decided to begin apartment hunting with the help of an American friend of hers. After two days, we found one in a nice neighborhood near Sepulveda Boulevard, within walking distance of a laundromat and two grocery stores. It was also on the bus line to different parts of the city, including UCLA. We agreed it was ideal and each signed a separate lease, splitting the rent in half. Luckily, the rent included water, leaving us to pay the electricity and telephone bills separately, for which we devised a system. We also agreed to contribute $60 apiece each month for food. This money was put in a safety deposit box in the apartment and when we needed groceries, we would withdraw the necessary money from the box. Each expenditure was matched by receipts.

For about three months, everything went well. After that, problems started to crop up. My roommate changed the rules we had established for the food money. She borrowed cash from our savings and bought personal items. She also held impromptu dinner parties when I was not home and bought beer with the grocery money. We found ourselves with no money by the middle of the month, which was a serious hardship because we did not get paid for our teaching jobs until the end of the month. I was frustrated but confronting her only made matters worse. She locked herself in her room and turned on her stereo, making it impossible to read or do homework at home. I got some relief over the weekend when she left the apartment and spent time with her friends. At the end of the month, I refused to contribute to food by letting her know that, from then on, we would each be on our own. This was the best defense because two days later she told me she was moving out to live with another South African who had just arrived. I knew that it was going to be difficult for me to pay the entire rent on my own, but I was ready to figure something out. I decided to talk to the property owner about the problem, and I was surprised that he was understanding. He allowed me to pay my share for that month while I looked for a replacement.

At one of our department gatherings, I announced I was looking for a roommate, requesting help in spreading the word. One of my classmates asked me later if I would consider her as my new roommate. She was sharing a two-bedroom apartment with three other people, and she wanted to have her own bedroom. I was happy about finding a solution so soon, so I agreed. With my earlier experience, I decided I was not going to engage in a sharing plan. We agreed on how to pay the phone and electric bills and keep grocery bills separate.

Everything started off well until cultural and religious differences came into play. My roommate was Jewish, and I knew nothing about Jewish culture or religious restrictions. One day I came back from school and found all but two kitchen cabinets taped. The utensils I had in the apartment were squeezed into the space she had designated for me, with a note on the counter informing me that I needed to observe kosher restrictions. The note explained that because I was not Jewish nor a vegetarian, there was a problem with keeping the kitchen kosher. A kosher kitchen, she added, must have different sets of utensils, one for meat and poultry and the other for dairy foods.

I took the note and sat on the couch in the living room to think while trying not to be outraged. I could not understand why she chose this form of communication instead of a face-to-face dialogue where both of us could produce a workable solution. I was now faced with the dilemma of how to create an environment where this and other differences could be settled amiably. I decided to take time to cool off and gave myself until the next day to consult with my friends at school. That evening, my roommate and I did not see each other, and in the morning, I left before she awoke. At lunch that day, I conferred with two of my trusted American friends in the same class. They were confused and did not have any advice about how to resolve the conflict. After a week, my roommate decided to move in with her boyfriend while I looked for someone else. However, this experience made me doubtful about finding a suitable roommate. I decided to look for a one-bedroom apartment.

In addition to looking at newspaper ads, I asked friends to help with the search.

One day at church, I met a woman from Uganda. We started conversing about home, the political trouble in her country, and her family. She was a nun before she came to America. The Red Cross helped her escape the regime of Idi Amin, the Ugandan military officer and president, when they attacked her convent and many members of her household. First, she escaped to Kenya and from there was granted passage to the United States. When I told her about my search for a one-bedroom apartment, she rescued me. Since she had just gotten married and was moving into her husband's house, she would be leaving her current apartment. She agreed to introduce me to the property owner and put in a good word for me. We agreed to meet the owner a day before her lease ended. The property owner had a good relationship with my newfound friend, and I thought that would buy me some favors. However, after the meeting, she did not offer the apartment to me but promised to call me by the end of the day to give me her answer. I was anxious because I liked the apartment itself and its location on the intersection of two main roads in the area, Sepulveda Boulevard and

Venice Boulevard. The neighborhood was full of small mom-and-pop shops, mostly Mexican, Cuban, and Asian. The intersection had more choices for buses to Venice Beach, Marina Del Ray, Santa Monica, UCLA, and to major grocery stores. I was excited about the possibility of securing an apartment that would finally give me the freedom and independence I desperately needed.

I did not get a call until around eight o'clock that night. When the phone rang, I picked it up with anticipation. I heard the landlady's voice at the other end of the line. She proceeded to let me know she had decided not to lease the apartment to me. I was devastated but tried to hide it and thanked her for the call. But I also told her that if she changed her mind or had another apartment become available, I would love to rent it. She did not say anything, and we hung up. I called my Ugandan friend and told her what had happened. She was surprised and said she was sorry. She promised to keep her eyes and ears open and would let me know of any opportunities. I went to bed resolved to continue looking and not to give up so easily.

The next day, I went to school as usual, but instead of going straight there, I took a bus to Santa Monica to check out a place I had seen advertised in the *Bruin,* UCLA's student newspaper. An apartment in Santa Monica would have been ideal because there was rent control while other places raised their rent by two percent each year. The place I went to check out was not ideal because it was an apartment on the ground floor and near the waterfront, and I saw quite a few homeless people camping near the corner of the building. I felt unsafe and worried about coming home late at night or being away from the apartment for an extended period.

As I was riding the bus back to my apartment, I prayed that I would find a place soon because the end of the month was near. As I was entering the apartment, I heard the phone ringing. I took my time getting to it because I was not expecting a call from anyone, especially around three in the afternoon. I picked up the phone and answered half-heartedly because I was tired, depressed, and hungry. My energy picked back up when I realized it was the property owner calling me. She told me that the person who had been interested in the apartment changed his mind and if I still wanted it, it was mine. I wanted to scream with joy, but I held myself back and politely thanked her for reconsidering me. I was so happy and did not feel tired or hungry anymore. I told my current landlord I would be moving at the end of the month and thanked him for the time I had been at his property. That evening, I did not see my roommate because I retired to my room early, so I simply left her a note on the coffee table.

Surviving Graduate School in California

Graduate school was rigorous, but my experience in England was instrumental in my ability to navigate the academic terrain at UCLA. I wanted to spend as little time as I could at UCLA, so my target was four years. I knew I would have to complete my teaching responsibilities and my required graduate classes. I was lucky because I was admitted into the Ph.D. program, which allowed me to avoid a few courses that are deemed remedial to ensure that a student is ready to advance to candidacy. Since I already had two master's degrees, one from Tanzania and another from England, and my friends were only starting their first master's, they had more tasks to complete. I found out at the beginning of my first year that I had only a few required courses, which I was able to complete that year. I was left with several electives in the second year and dissertation research in the third. I was convinced that if I played my cards well, I should be able to use my fourth year to write and defend my dissertation in time to graduate the same year. I made sure that the classes I took in my third year would contribute to my research ideas for my thesis, while I tailored the term papers to fit specific chapters of it. My teaching load was reasonable—one beginner Swahili class, just like in England, which made my work manageable.

There were two highlights during these four busy years at UCLA that remain memorable. One was in 1984 when I was nominated to receive a distinguished graduate teaching assistant award. This was my third year in the graduate program, and it came to me as a shock because I did not even know such an award existed let alone the process for selection. It was also interesting because a couple of months earlier, I had dared to seek an appointment with the dean to request a scholarship upgrade. My argument was that the amount I was given was too little to allow me to save enough for the three summer months when I did not receive any financial support, which forced me to find a summer job. Unlike the other students, I could not afford to go back home to cut costs or take temporary employment since I did not have a permit to work off campus. The dean tried to convince me that the system was set up that way and there was nothing he could do. I kept insisting that he could find money somewhere to offset my expenses for books and other supplies, which would help greatly. Realizing I was not going to leave without a positive answer, he decided to create a new allocation for me, an additional $250 a month. This was huge because it was equivalent to half of my rent, and I planned to save all of it for summer expenses. So, when the news came that I was nominated for an award, I thought my lucky stars were still shining brightly.

There was a graduate committee that nominated several graduate teaching assistants (GTAs) and then students were asked by the graduate coordinator to send letters of support to the committee for the candidates. The selection required end-of-term evaluations from the students taught by the nominated GTA and the GTA's academic standings. For the 1984 nominations, I emerged as the winner. The award came with $15,000 toward the completion of a Ph.D. dissertation and $400 in an envelope presented at the award ceremony. I was in heaven when I got this surprising news. My students were the happiest. They threw a surprise party for me the following weekend that started at six o'clock and lasted into the wee hours of the morning. That is something I learned about California; they know how to party. I appreciated my students for recognizing my efforts to help them learn and enjoy my language.

The second highlight was also in 1984 during the Olympics. UCLA was one of the venues. Since it was summer and I could work on campus, I applied for a position in security. The first assignment was as a security guard from ten o'clock at night to six in the morning. This was the hardest job I have ever done. We were required to scout our designated areas at half-hour intervals and take a ten-minute break from time to time while standing up. I did not realize one could fall asleep standing up. After one week, I decided to request a shift change to work during the day. I was relieved when my request was granted. My new post was to monitor the metal detectors at the entrance of the UCLA Olympic village. This was an exciting position because I got to see everyone who entered the village. I met many athletes from different countries and collected a lot of different Olympic pins and spoons that I still have.

Meeting athletes from Tanzania was interesting. I startled some of them when I greeted them in Swahili. One of them asked me if I was sent by the Tanzania government to watch them. I was amused by this but assured them that I was just a student trying to earn some money and be a part of this great event. One of them gave me $100 and added, "Go buy yourself books for your studies." This was a lot of money, and I thought that was a very generous and thoughtful gift from a total stranger. Unfortunately, he did not come through the gate again during my shift and I did not get to ask him his name or what event he was competing in.

As a graduate student, I also did several odd jobs on campus. The annual (nine months) GTA scholarship was about $10,000. With my one-bedroom apartment, I did not have much left after paying for rent, utilities, and groceries. For a long time, I did not have a car and walked everywhere, especially on weekends because the city buses did not run in my neighborhood. On weekdays, buses did not run after six, so I made sure I left campus before then.

Walking to church or the grocery store, I passed many parking lots and car dealerships. I loved picking out my dream car and imagined how I would look behind the wheel. I did not think I would be in the US long enough to make enough money to buy a car. Owning and driving a car in the United States was wishful thinking, I thought. It would be a miracle if it happened.

Without a car and because there was no weekend bus service in my neighborhood, grocery shopping was always a challenge. I walked to the grocery store, and I carried groceries in a big manila basket balanced on my head without holding it. This freed both of my hands to carry small bags in each hand. This was from experience. African children as young as five learn how to carry loads of various weights on their head. In the villages, mothers can be seen carrying heavy loads on their heads, a baby on their back, and an item in each of their hands while they walk briskly to and from the market or farm. Practice makes perfect and experience turns into a skill.

For pedestrians and drivers in passing cars, this was a spectacle. They stopped to watch me, and the drivers honked. One day I told my American friends in Palo Alto, Jean and Charles, about these experiences. I pantomimed my story, causing them to laugh hysterically. I laughed with them to assure them it was fine to laugh.

They said they were sorry I was having such a hard time. I tried to assure them it was not much of a problem, but they did not think my story was just an amusing incident but a cry for help and a problem that needed an immediate solution. Later, I was surprised to learn that my funny story created the miracle I had been hoping for. I received a card from Jean and Charles at the end of that week. It had words of encouragement and a check for $2,000. The note ended with: "Look for a nice used car." I was dumbfounded by this kind and generous gift from friends I had met in Tanzania five years ago, and who considered me a family friend. This gesture was beyond being a friend. They had taken me under their wing as one of their own. I knew they were pleased that I was admitted to UCLA for graduate work, but I did not realize they had extended themselves to being my guardians in America, sixteen-thousand miles away from my birth home. I found myself shedding tears of joy and wondering how to thank them or what to say to them. Returning the check would have been an insult, so I decided I would have to find a way, someday, to thank them for their generosity.

For the remainder of my time at UCLA, Jean and Charles called me on the phone often or sent a letter to check on how I was doing. Occasionally, I would receive a check with a letter telling me to stop eating junk and buy good food. They also sent me a plane ticket each Thanksgiving and Christmas break to spend time with their family. I always had a gift under their Christmas tree, just like everyone else in the family.

When I got the car money, I felt like an American teenager getting their first car just because they turned sixteen. The only difference was that I was not sixteen and I did not expect it. So, this was truly a major, generous gift, a gesture that told me they considered me a beloved family member. Like typical American parents, they extended support when their children needed it, this time securing a means of transportation. From that day on, I referred to them as "mom and dad."

I had some friends in my neighborhood who I would often spend time with. I asked them to help me find a used car to buy. They advised me to look in the advertisement section of the newspaper and not the dealership because the latter inflated their prices. I was a woman on a mission to find the car I had been dreaming about but which I thought was out of reach. After two days of looking, I saw an ad for a car I liked. I called my friends and one of them agreed to drive me to meet the owner. The seller was an elderly woman who was giving up driving. The car was like new. She did not drive it much, so it was in mint condition. She wanted $2,500 for it. We started talking and she detected my accent. She asked me where I was from, and when I said Tanzania, she noted her admiration of the president, at that time Julius Nyerere. She thought he was a devout man who cared deeply for his people more than his personal wealth. I did not realize that she was very politically savvy and knew about many world leaders. We talked politics for almost half an hour, and she enjoyed every minute of it. She then asked me what I was doing, and I told her I was a graduate student at UCLA.

"Good for you. You must be very smart," she said. "And you came all the way from Tanzania? I am impressed."

"Yes, I came from Tanzania via England. I went to York University, but I was unable to finish there because the government withdrew my scholarship," I replied.

"Damn them Brits," she retorted.

"It is OK," I replied. "UCLA came to the rescue."

She changed the subject back to the car-selling business.

"Now that I know a little bit about you, my dear, I can sell my car to you. This car is like my baby. I know it is going to good hands," she said.

"Thank you, I will take very good care of it," I responded.

Then what she said next was a big surprise.

"OK, how about this? You give me $500, the rest you can use to help you at school."

At first, I did not think I heard her correctly. I put my hand on my mouth, totally dumbfounded.

"You go on, come back with $500 when you have it, and you can have the car. I will get the papers ready for you by the time you return. Give me a call before you come," she added.

"Thank you very much. I do not know what to say," I replied.

"You do not have to say anything. Go on, go back to school," she called out as she left the driveway where the car was and entered her house.

My friend and I were in disbelief. We drove away with a plan to come back after a day or two to give the nice woman time to get whatever papers she needed to put together. I could not believe that I would have leftover money from the Praels' gift to use for insurance, registration, and any other expenses associated with the car. That evening I called Jean and Charles to give them the news. They did not believe I was that lucky. They were worried that there was something wrong with the car. They wanted me to find a good mechanic to inspect it and do a road test to make sure it was safe to drive. I promised them I would do that.

Because I already had a learner's permit, I decided to go to the Department of Motor Vehicles to make sure it was in good standing. Also, I found a driving school to learn the American driving rules. While in America we drive on the right, in Tanzania and many of the British colonies, the driving is on the left.

I felt I had covered all the bases. I asked my friend to take me back to pick up the car. My friend's husband found a mechanic for us who owned a repair shop. He agreed to come with us to pick up the car. Once we got there, the woman came to meet us and handed me the yellow registration card after I gave her the money. She also gave me a file with copies of the car maintenance record over the years. My friend drove off by herself and I drove in my new car with the mechanic to his shop to do a full diagnostic test.

At the garage, the mechanic went through the file and was impressed by how well the car had been maintained. It had only twenty-thousand miles on it over ten years. Because it was Saturday, I did not have anything important on my agenda and could wait for him to run all the needed tests. The only thing he did was change the oil and clean the filter. The car was good to go. My task was to finish my training, take the driver's test, and get on the road. I gave the good news to Jean and Charles that evening, and they were happy. I also sent them a copy of the history and service report from the mechanic. Knowing they were comfortable with me being behind the wheel of this car gave me some peace of mind. After church the next day, I went to the store and bought two thank you cards, one for Jean and Charles and the other for the owner of the car. It was important to thank her because this was truly a gift to a total stranger she had met for only half an hour before making her decision. To me, this was incredibly generous and exceptional. I was beginning to believe in the California I had imagined in high school.

When Jean and Charles saw the reports, they sent me a card with congratulations. It was a special card because it had a picture of a woman in a fancy car. My two-door Honda Civic was fancy to me. This was a perfect message. They always expressed their pleasure at my accomplishments, something that made me work harder because I did not want to let them down. Every time I had some success, such as my teaching award at UCLA, finishing my Ph.D., getting a job at Stanford University right after I graduated, moving to the University of Georgia (UGA) to take a tenure-track job, buying my first house in Georgia, a promotion to an associate professor, getting my green card, and securing multiple grants for a variety of projects at UGA, I shared it with them and I received a congratulations card. When I look back, I always wish Charles had been alive when I became a US citizen; he died in 1997. Jean, his wife, died nineteen years later in 2016. When I became a US citizen, she invited me to Seattle, where she had relocated, to celebrate with other family members. During the years after her husband's death, Jean continued the tradition of sending a congratulations card for every achievement I attained such as the promotion to professor, being recognized for my contribution to the University of Georgia, and being named "University Professor, an award given to one professor a year by the University System of Georgia." They will always be my adoptive parents in America; they created a family for me away from my ancestral home.

Back to my prized commodity, my first car in America. Having this car meant a lot to me. When I got the car, I started thinking of the days when I walked everywhere, looking at cars in parking lots and picking out the one I wished I could have. But it was an adventure when I was behind the wheel. In the beginning, I had to remind myself that I should keep on the right side of the road because in Tanzania we keep left. I cannot forget one incident during a driving lesson. I got into trouble when the driving coach asked me to make a turn onto a side road. As soon as I made it, I went all the way to the left side of the road. He did not stop the car with the controls on his side but looked at me and asked, "And what are you doing now?"

Not knowing why he was asking me this question, I responded, "Driving, of course."

"On which side of the road?" he continued.

That startled me, "Oh, oh, I am sorry. I should be on the other side of the road," I replied.

"Okay, then do it." I quickly moved to the right. Then he added, "If you do this, you will cause an accident before you know it and you will be in multiple problems, including losing your license, if you ever get it."

The final part of his statement, "if you ever get it," made me doubt myself. I thought he did not think I should be driving. From there on, I behaved myself until I went for the driving test. I did well on the test except for parallel parking. My tester was sympathetic and passed me anyway. When I came back to my driving coach, I was beaming. He asked me how I did, and I told him that I passed. He had an

"I can't believe it" look. I am sure he was concerned that I would forget again and drive on the wrong side of the road. He let me drive his teaching car to my apartment and then congratulated me for passing the test.

Now that the training and test were done, I could venture out driving alone. I started by driving to school, the grocery store, and church. I gave friends rides and began gaining confidence by the day. While I enjoyed being behind the wheel, I was terrified of the highways and tried to avoid them at all costs. One day my friends and I were coming back from school. It was around two in the afternoon and traffic was moderate. Before I turned to drop off one of my friends at her apartment, a request was made to stop at Kentucky Fried Chicken. The location was at the intersection of the main thoroughfare, Culver City Road, and an entrance to the 405 freeway. Next to it was a gas station. I had never been in this area, so I did not how where the entrance was. My friend Jemma said, "Make a left over there."

I saw an entrance on my left, turned on my signal, and then quickly turned in front of approaching traffic. Lo and behold, this was the entrance to the freeway and not Kentucky Fried Chicken. Jemma shouted, "I meant the following entrance. You need to keep moving."

"Hell, no," I replied. "I am not going on the freeway."

"You cannot stop!" everyone in the car shouted.

"No way," I said, then pulled to the side of the road, and put on my brakes.

"What are you doing?" everyone asked in unison.

"I…am…not…going…on the freeway," I declared.

By now cars had lined up behind me. I told Jemma to get out of the car and ask the other drivers to let me reverse from the freeway entrance. She did. She talked to the driver right behind me. It was a young man who decided to help. He got out of his car, went to the entrance, and asked all the drivers to wait while he guided me out in reverse. Once close to the main road, he acted like a police officer and stopped all the oncoming cars until I was safely inside the Kentucky Fried Chicken entrance and off the main road. I sighed with relief and thanked my lucky stars that there were no police in the area. I would have gotten a ticket. I vowed never to take directions from anyone again and always plan my trip ahead of time.

— 12 —

Where There is a Will, there is a Way

People believe that graduate students are the lowest paid at institutions of higher education. I thought so too during my time as a grad student and constantly thought of ways to increase my purchasing power and to save enough for trips, particularly to go home during the long summer vacations. I took various temporary jobs on campus and chose shifts that suited my teaching and academic responsibilities. This proved to be an excellent way to supplement the money I received from my teaching fellowship. But I also realized that having a car was like having an extra mouth to feed. It needed its own budget. I applied for a position at the food court on north campus. I was successful, and this was a great location, close to my department and classrooms. At the beginning, my job title was "dishwasher." The food court had a dishwashing machine, but the management required that we rinse the plates and utensils before we put them on the conveyor belt that took them to the washing machine. The bigger pots had to be hand-washed. I hated washing the big pots, but since I wanted the money, I did not complain. I had a supervisor who inspected the pots and, if satisfied, he would allow me to leave early. I never had issues with my assignments. I showed up on time and left when all my work was completed. And employees could have one free meal during each shift, which was a wonderful way to save on groceries. I opted for takeout, and because the portions were big, one meal was enough for two servings.

After a month of dishwashing, I was promoted to chef's assistant. I assumed the early morning shift to start the breakfast. This meant arriving at the food court at six o'clock in the morning. The head chef was always there by that time. I started the oven, arranged the bacon on the bake trays, and placed them in the oven. Then I filled the serving trays on the counter with an assortment of breakfast foods as directed by the head chef. At about seven-thirty, the chef allowed me to have my breakfast, and then I left for class. I taught from eight o'clock to eight-fifty in the morning. Because I did not have classes of my own while I worked on my thesis, I had additional hours to put in at the food court. I always stayed until the afternoon because the management needed more workers during lunchtime. For the lunch shift, my major responsibilities were preparing condiments for salads, sandwiches, and burgers. I sliced different types of cheeses, onions, lettuce, and tomatoes, and grated cheese for a variety of uses. I also ground the meat that was used to make meatballs for spaghetti sauces. One of the foods I never ate at the food court was spaghetti with meatballs. This was because we made a lot, and if we had leftovers, the chef would keep it in the freezer room to serve again the next day. To prepare the leftover spaghetti, the chef boiled water in a big pot, put the cold spaghetti into the pot, and let it sit for a while before it was put into the serving trays. We always prepared new meatballs because there were never leftovers. Nevertheless, I disregarded the spaghetti and opted for chicken and French fries. I liked the chef and learned a lot from him, especially about the dishes that were foreign to my culture.

In a similar way, teaching at UCLA provided a new reality, different from the British culture I had experienced in England and had become accustomed to from middle school through high school and college preparatory school. The system was more teacher-oriented where students depended solely on their instructor for knowledge and success was rooted in passing exams. If students did not pass their tests, the teacher shared the blame.

At UCLA, I had to learn to cultivate a different teaching philosophy and perspective. I worked hard to let go of my old teaching habits, but they showed up at times. I still considered myself the custodian of the knowledge I was imparting to the students. I expected the students to sit still, listen, and then use ten minutes at the end of the class to do a summary of what I had taught that day. Participation was mostly through questions and answers and, in a language class, occasional group work. Many students did not like group work. I did not know why until later in my teaching experience. It was easier for a student to stay quiet and let those who were fast learners answer all the questions when the teacher posed them. What students hated most was feeling inadequate in a group setting. I was also surprised to find out that students did not like to be called on in class to answer questions; it was mostly on a volunteer basis. Calling on a student to speak in class when they have not volunteered was considered "picking on the student." Once I learned that, I modified my teaching approach, but this created another problem that did not become apparent until after a midterm. One of my Black students did not do well in the midterm. I had expected this to happen because she missed a lot of classes, and she did not speak much in class, even in group settings. She was one of those students you could describe as extremely introverted. After distributing the graded tests to the students, I noticed that the student was not happy with her grade. I had written "see me" on her paper hoping she would come during my office hours. I wanted to explore with her how I could help because I wanted her to do better in the finals. However, the student left the classroom before class ended. Later that afternoon, I was called to the main office to meet with my department head. Little did I know that this was related to the student. After leaving class, she went straight to the dean's office and claimed that I discriminated against the Black students in my class. The dean thought I was a white instructor and referred the case to the department head before it was sent to the Equal Employment Opportunity Commission (EEOC).

When the department head gave me this information from the dean, I looked shocked. It never crossed my mind that I could be considered a discriminator. I knew I was not racist either. I gained a new perspective about the term. I was not sure how to respond to the accusation but explained that I did not think I was guilty of the charge. I also told the department head what I thought was going on with the student. The department head was understanding and asked me to observe the situation and report back in a week. The student did not show up in class for the rest of that week. Later, I received notification from the registrar that she had dropped all her classes for the term. To avoid future misunderstandings, I made it a rule to meet with each student at least once during the term, and more with those I thought were struggling. I also introduced cultural awareness as a special aspect of the class. This involved the students working in groups, selecting an aspect of Swahili culture, studying it, and then presenting it. The goal was to cultivate cross-cultural understanding and to encourage the students to develop a cooperative rather than individualized learning process. The once-a-term gathering required the students to work in groups of four or five to prepare an African dish to share. I provided each group with a recipe and asked them to do a trial run before they made the dish that would be shared. The gathering was scheduled on a Saturday afternoon at my house or at a park of their choice. This initiative became popular and something to look forward to each term. Because of its success, I made it a regular event for all my classes in subsequent years. It was interesting to see students who had graduated but were still in the area come to the gathering while students in the class invited their friends and family members. It was something they looked forward to, and when the groups grew too large to accommodate at my house, we looked for space on campus. The event allowed me to teach about Africa.

In addition to my work as a teacher, my work as a student at UCLA went well the first two years. I intended to complete my degree in four years. I was in a hurry because I had spent two years in England hoping to complete a Ph.D. I did not think time was on my side, and I wanted to return to the workforce. After two and half years of the necessary coursework, I developed my research and writing plan. I knew I was challenged in writing because I did not own a computer and, unlike now, there was no public access to computers on campus. To get my work typed, I needed to hire someone, but the cost was astronomical. A friend of mine who owned a stationary shop in Santa Monica, and who had a computer, offered the space and the shop computer if I wanted to do it myself. I accepted the offer knowing that I had to learn how to use the computer as well as schedule myself around the time she had available to use it. She agreed to give me a two-hour crash course. In those years, computers were not as simple as they are today. To type anything printable, you had to include a lot of symbols that would program the computer to translate it to printable text. It was computer programming. Work was not saved on an easy-to-carry thumb drive but a larger and bulky zip drive.

To make sure my friend and I had uninterrupted time on the computer, she gave me a key to the shop so I could go in when she was not there. Because I was still teaching, doing research, and working at the coffee shop on campus from 6:30 to 7:30 a.m. before I went to teach my 8 a.m. class, I scheduled myself to be at the shop from 4 to 6 a.m. Monday through Friday. I had additional time on Sunday because the shop was closed. With that schedule, I was able to implement the work plan I had put together, which allowed me to complete my first draft in three months before I surrendered it to an editor. It was much easier after the editor's remarks to clean it up and submit it to my supervisor for her first evaluation.

Looking back, I realize that where there is a will there is a way. Getting up at three o'clock in the morning and driving to Santa Monica was an effort. It was also dangerous to leave my car outside of a deserted shop with the lights on. While I did not think it was dangerous at the time, I suppose it could be riskier now.

The rigid schedule I had made for myself enabled me to complete writing, editing, and submitting the dissertation to my committee in good time to make graduation in my fourth year. Some of my committee members thought I was rushing and reminded me I had at least six years to finish. There were students ahead of me who had maxed out their time and some were even nearing twelve years. Still, nobody was rushing them. However, I was not planning to do that. I wanted to finish and leave. The encouragement came from my major professor who worked with me every step of the way to realize my goals. She was successful in convincing one of the committee members of my hard work. Eventually, I was cleared to graduate as planned in June 1985.

With a diploma in hand, I decided to take advantage of an opportunity granted to international students to participate in a program known as Optional Practical Training (OPT). To participate, one had to apply through the Office of International Studies. The application was sent to the U.S Immigration and Naturalization Service (INS), and if approved, one had up to a year to complete the program. At that time, things were not as complicated as now, and it took two to three weeks to process the application. I was one of the lucky ones. My application was granted within two weeks. With the necessary paperwork secured, I started applying for teaching positions at various schools where I thought I would be a good fit.

I liked Southern California a lot and was not inclined to relocate at this time. If possible, I wanted to avoid Los Angeles because of the traffic and pollution. I liked my college town, Westwood, and the surrounding communities of Brentwood, Hollywood, Culver City, Santa Monica, and Venice. The way the neighborhoods were laid out created a calming effect, which heightened the sense of peace and security. I could also continue to stay in my apartment if I found something in the neighborhood. My favorite pastime was driving around different neighborhoods and admiring the houses of the wealthy people who lived in places like Brentwood and Hollywood. However, I knew there was no chance in hell I would find OPT in these neighborhoods. I needed to search outside my comfort zone.

Some friends introduced me to school districts and community colleges where I could explore such an opportunity. The first high school that invited me for an interview was a private Christian school, thirty minutes from where I lived. The principal showed an interest in hiring me to teach English but did not make an outright commitment. I was to hear from her in about a week. While I was waiting, I received a phone call from a community college in the Los Angeles area to visit the school. The phone call came on a Wednesday, and I was asked to be at the college around 10 a.m. the following Friday. I thought I was going for an interview, so I prepared well for it. I still had my little Honda Civic, but I was terrified by the thought of driving to this college because it required me to take the 405 freeway. I realized that if I wanted this job, I had to tough it out and bear the freeway congestion. I psyched myself up, went to bed early on Thursday night, and planned to set out around five-thirty in the morning. I thought this would put me ahead of the traffic. Little did I know that the 405 freeway does not have a light traffic time. It was like people were always on the go on this highway. Leaving early that morning proved to be a wise decision because I needed a lot of time to get to the college, which was an hour away from where I lived. There was no GPS at the time, so I secured a map and planned my trip by highlighting the exits and the streets to the campus. I told myself that the job itself did not look that hard if I could survive the 405.

The trip took longer than I thought because of traffic congestion on the highway. I arrived a little after nine-fifteen and had plenty of time to relax in my car and to find the Department of Languages. I had a thermos full of tea, a couple of pieces of toast, and some fruit. I decided to have my breakfast before I went to what I thought was an interview for a teaching job. I headed to the location five minutes before the meeting. When I arrived, I found people already seated. After exchanging greetings, I introduced myself and stated my reason for being there. One person stood up and introduced himself as Dr. Smith and welcomed me to a chair next to his. He was knowledgeable about Tanzania; he had been there several times. He also taught Swahili at this college and was hoping I would take over his classes so he could teach English only. We had a little chat before the room filled up with other faculty members. I started to worry about the format of the interview because I had had no warning about what would take place. What followed was a big surprise.

A woman sitting at the head of the big conference table called the gathering to order. I learned later she was the department head. She asked her secretary to read the minutes of a previous departmental meeting. After the usual formalities, Mr. Smith raised his hand and said, "I would like to acknowledge the presence of Dr. Moshi from UCLA. She was invited to meet us today." The department head welcomed me and added, "We are happy to have you here with us." Then the meeting continued as planned. I learned that this was the English Department, which also oversaw the teaching of foreign languages (Russian, French, Spanish, and Swahili). Swahili was introduced by Mr. Smith, who had a passion for it, and it was his linguistics research language. Like many departments on college campuses, there was no shortage of disagreements, some legitimate, some petty. The arguments often were sparked by trivial comments from a faculty member or the chair. Then the temperature rose as new issues were brought to the table. The new semester was just beginning, so there were a lot of issues to be discussed: the curriculum, teaching schedules, department assignments, and student advisement. Since classes were starting Monday, this meeting was necessary to put everything in order before the students arrived. The differences were precipitated by personal conflicts among the staff as well as outright disenchantment with the department head. She seemed well prepared for this crowd and held her ground. At one point, one of the faculty members threw a book he was reading at her. It missed hitting her square in the face. This qualified as an outright assault but nobody said or did anything about this behavior. The meeting continued as if nothing had happened. By now, I was feeling extremely uncomfortable and out of place, especially because I did not know why I was at this meeting.

Once things calmed down a bit, the department head welcomed me again and proceeded to assign me classes. This was Friday and I was expected to start teaching the following Monday. I had three classes to teach, two English 101 courses and one Swahili class. The English classes were scheduled for eight and ten in the morning and the Swahili class was at four-thirty in the afternoon. I was in shock when she announced the schedule. Several thoughts were swirling in my mind. I did not know I was hired because no offer letter had been extended to me. I was also concerned by the schedule starting in three days because I did not live in the area and had to drive at least three hours to get there due to traffic. I thought making an eight o'clock class was nearly impossible unless I left my apartment by 4 a.m. Arriving so early and finishing classes around five-thirty was more than I was willing to do. I would be getting home no earlier than nine o'clock at night because of the traffic. The schedule was also for five days a week, Monday through Friday. Without an offer letter, I did not know what my salary was, whether I had an office, or where the classrooms were located. I felt like a fish out of the water. The atmosphere in the department was also a concern, if not outright disturbing. The department head continued to inform me that the department was moving to another building and described its location on campus. I asked how I was to locate all these places, the classrooms, and the students I was supposed to teach on Monday morning. She answered, "You will find out, one way or the other."

While I was still stunned, the department head ended the meeting, and everyone started to leave the room. It was around noon when Mr. Smith touched my shoulder and said, "Do not worry, it is not as bad as it looks. I am sure all your classes will be full, especially the two English classes. For those, you will be done teaching by noon each day. Your Swahili class roster will also show a full class, but you will never see any of these students."

I was taken aback, and asked, "How does it work teaching students I do not get to meet?"

"Oh, you just give them a grade at the end of the semester. They are all football players, and they have no time to come to class," he said.

"How do you give them credit when they have not done the work?" I asked.

"Well, well, Dr. Moshi, this is how we have always done it. If you want this job, then that is what you need to do," he replied.

I thanked him for talking to me and I left the building for the parking lot. I turned around and went back to the department to see if I could get more information about where to go on Monday. The secretary walked me to the new office building and assured me that she would be there by the time I arrived to show me the classroom. I thanked her and went back to my car. I got in, still in disbelief about what I just witnessed and heard. I started my car and headed back to the 405 freeway toward Culver City and my home, sweet home. Traffic was out of this world at that time. My car did not have air conditioning and, naturally, in the middle of summer, it was extremely hot inside. I tried to open all the windows, but it did not help. I started thinking seriously about this job and whether I really wanted it under those conditions. I said aloud, "God, find me a good job, and do not let me come back this way again."

With the heavy traffic, I got home around three-thirty in the afternoon, and I was both tired and distraught. After a little snack, I decided to go for a walk to clear my head. By the time I came back to my apartment, I had a plan. I was not going to give up the opportunity, but since I did not have a signed contract for the job, I needed to request it before I started teaching any class. After that, I planned to request time to relocate to the neighborhood to avoid the long drive back and forth to Culver City. That decision made me feel better.

All weekend, I was preoccupied with what I was getting myself into. I grew more and more convinced that this was not the right place for me, but since I had nothing better, it was time to be wise and contemplative. I told myself that I would not take Dr. Smith's advice about the Swahili class and that I would demand that all the students enrolled attend if they wanted to get a grade. I was determined to locate their counselors, and without disclosing what was irritating me, impress upon them what I expected from the students under their advisement. I was going to follow the tradition I had appreciated at UCLA where a student had a certain number of excused absences. This included allowing student-athletes to miss class if they had an away game and had to travel out of town. I also would provide support to the students' tutors to ensure they were not left behind and would be available during office hours and for requested appointments to assist any student who needed extra help. I was not sure if this was going to work out as I had planned but was convinced I had an advantage as a new instructor because I could play innocent.

Saturday and Sunday came and went quickly. I did not want to invest all my time and energy on this potential job, so I decided to do the minimum and prepare introductory remarks for all three classes and a "get to know you" game. Sunday afternoon, I spoke with a friend who was a lawyer about my dilemma. She was shocked at what happened on my first visit to the school and advised me not to show up on campus, let alone in any class, before I received an official offer letter. She also indicated that the administration would not come to my defense should anything happen in any of my classes because I had not filled out the appropriate legal paperwork to allow me to teach at this college. All this confirmed what I had been thinking. I realized that my eagerness to find an institution to complete my OPT might lead me down a cliff. With her advice, I decided that

I would call the department first thing Monday morning and express my hesitation to start teaching without a letter of employment.

By eight on Monday, I was ready to make the call. The department head was not at the office, but I spoke with the office administrator who agreed to take a message. I made it clear that I would not teach any class before I received an offer letter because I knew it was illegal. She did not comment but noted the department head might call before the end of the day.

I did not receive a call that day. However, I did get an interesting call from the director of the Stanford/Berkeley Joint African Studies Program. They had received federal funding that required the establishment of an African language to strengthen their certificate program. Someone had told them that I had just graduated from UCLA, and they wanted to know if I would like to join their team and be a part of the Linguistics Department. I thought God had answered my prayers. I did not want to sound too eager, although I was excited about the possibility to teach at such a prestigious school. I indicated that I would think about their offer and let them know in a couple of days. Wednesday could not come fast enough. I wanted to let them know I would be happy to join them.

Wednesday came and I still had not heard from the other institution. However, this did not matter because I had Stanford on my mind. Around noon, I called Stanford and told the director I had decided to accept the invitation to teach there. I was not sure who was more excited about this decision. I had a feeling the director had been waiting for my call because he sounded happier than me. I thought he might have been worried that I would say no. During this time, I was afraid they might have reached out to someone else. But his reaction assured me that all was well, and we were both on the same wavelength. He told me the Linguistics Department administrator would be calling me with further information. In my mind, I had already started preparing for a move to northern California.

With the decision to go to Stanford, I planned to call the other job early Thursday morning to let them know I had changed my mind and would not be joining their department. I did not speak to the department head but to the administrator who told me that the letter had not been written. Somehow, her statement relieved me because there was no commitment from them, and I would not commit to them either.

Later that Thursday, the administrative secretary at Stanford called and said she was sending me a package with information about the department along with some forms I needed to fill out and mail back. Included in the package would be my letter of appointment. I was glad to hear that and felt I was finally back on the right path for wherever I was supposed to be going. The job at Stanford came at the right time. I was drowning and someone threw me a life ring. I needed to sail on to my next destination. Stanford was one of the schools I had applied to for graduate school, but UCLA had accepted me while Stanford offered me an opportunity to apply the following year. Now, Stanford was offering me a new beginning after graduation. This was indeed a life full of unknowns.

Preparing to move to the Bay Area did not take long. My brother Ladi had just arrived in California to start graduate school at UCLA. He was going to take over my apartment lease. So, I decided to leave everything behind except necessary personal items. I had enough savings to furnish my new place in the Stanford area. About three weeks later, I was packed and ready to leave Southern California. Ladi agreed to drive me and my belongings. We had agreed he would fly back to Los Angeles since I was keeping the car.

We left Culver City around six in the morning. We had planned to make two stops in Santa Barbara. The first was to visit a friend and colleague from the Linguistics Department at UCLA. She had graduated a year ahead of me and was teaching at the University of California, Santa Barbara. The second stop was to see my major professor, Dr. Sandra Thompson. Not only was she my supervisor but also a mentor and someone I admired throughout my studies at UCLA. She was an expert in sociolinguistics and specialized in discourse analysis, universal grammar, and Mandarin Chinese research. She was a major influence on my keen interest in discourse analysis, the main subject that I taught and wrote about throughout my teaching career. As her mentee, she was excited for my new position at Stanford.

After the two stops, we drove straight to Palo Alto, the home of Stanford University. Many people would have visited the area before relocating to survey and find a place to live. I did not do any of that because Palo Alto was like my second home in America. My American adoptive parents, Jean and Charles Prael, lived in the area, and I spent every major holiday at their house. I felt like I was going home, and presumptuous or not, I had security. Jean and Charles were expecting us and since their adult children were no longer at home, there was enough space for both Ladi and me for a couple of days while I hunted for an apartment.

Although Jean looked happy to see us, she was a bit withdrawn. I later discovered she was mourning. Her father had died that very morning and she was in the middle of preparing his funeral. Her father had been ill for some time, and she had moved him from Seattle to Palo Alto so she could monitor his medical treatment. He had been cremated and Jean was going to take the ashes to Seattle to the family burial site.

After dinner, Jean announced that she and Charles were leaving in the morning for Mendocino for a few days. They had a cabin in that area near the ocean, and it was fitting for her to take time to mourn her father. She allowed us to stay in the house until they returned. She left us plenty of food in the refrigerator and some emergency cash. This was Jean, always a mother who thought ahead for everything. This gesture was extremely touching, considering she had just lost her father, and beyond generous, for she was always thinking of others more than herself. This was the Jean I had known all these years, loving and caring even when she did not have the energy to do it. I was incredibly lucky to have her as my adoptive mother who provided me with a home away from home.

While Jean and Charles were gone, I visited the Stanford campus and continued the hunt for my new apartment. On campus, I visited the Linguistics Department and met the office manager, Gina Wein, who had become a friend through our telephone conversations before I arrived. I was surprised at how happy she was to see me. She hugged me like we were friends who had known each other a long time. The warm welcome assured me I was going to be happy at Stanford. As a fresh Ph.D. graduate starting a new job in an unfamiliar and prestigious place like Stanford, Gina's welcome meant a lot. I learned that Gina was a down-to-earth, very friendly person who cared for everyone without reservation, but she had a special spot in her heart for international students and colleagues. She often hosted spontaneous dinner parties at her house. Her husband was a great cook—I called him Master Chef because of his culinary skills. I had many meals at their home. Out of the blue, she would call and say, "What are you doing this evening?" I would reply, "Nothing much." Then she would immediately say, "Come for dinner." Knowing how good the food would be, I never passed up an invitation. When I thought I was going to be the only guest, I would be surprised to find the house full. That was Gina, her house was just as warm as her heart. She had one daughter, Tatiana, who was about seven or eight at the time, and a large extended family. Her mother and aunt loved me so much that they asked about me every day. I spent so much time with them that I ceased to be a guest but was a member of their extended family. Between Gina's family and my other adoptive family, Jean and Charles, America became more than a home away from home. A big thank you goes to Gina and her family. We have remained extremely close, and they are my second American family.

In terms of my immediate family, it was an asset to have my brother Ladi with me in the Stanford area. He received his master's degree in education at Stanford and had lived in the area for two years. While he assisted in my apartment search, extra help also came from Gina who pointed out places to look, simplifying our job even more. I found a nice, spacious apartment at the border of Palo Alto and East Palo Alto, a ten-minute drive to Stanford, five minutes to my adoptive parents' house, and fifteen minutes to Gina's house in Mountain View. I could easily walk to downtown Palo Alto, which had quite a few outdoor cafes and excellent bakeries.

With Gina's help, I was able to furnish my new apartment quickly too. By the time Jean and Charles returned, I had obtained most of the essentials and was ready to move in. Ladi and I stayed one more night after their return and the next day I moved into my apartment. He stayed a couple of days with me before he flew back to Los Angeles to prepare for his first year of grad school at UCLA. I had a couple of weeks left before school started, which allowed me to prepare my syllabus and get acclimated to the Palo Alto area. The department assigned me an office in the psychology building, adjacent to the Linguistics Department, where many of the faculty who also taught languages had offices.

Because I was hired to help Stanford develop and strengthen its African studies and languages programs, I had academic affiliations with the Stanford-Berkeley Joint African Studies Center. This was a Title VI center with funding from the federal government. Part of my salary came from this funding and all my language teaching materials were paid for by African Studies. This was exciting because I had the best of both worlds, associating with the African Studies faculty as well as the linguistics faculty. Of great interest was my involvement with Stanford's Center for the Study of Language and Information (CSLI), which did cutting-edge research in linguistics and had many professors with grants from the National Science Foundation (NSF) and International Business Machines Corp. (IBM) It was through this group that I received an opportunity to work with Professor Joan Bresnan who had a keen interest in African languages' contribution to linguistic theory. I was also introduced to IBM where I received my first grant to do language material development. This grant allowed me to travel to Kenya and Tanzania to collect video material that was used for my first online Swahili teaching material.

In addition to other world languages, Professor Bresnan was interested in Swahili, Chaga (my first language), and Chichewa through an association with Professor Sam Mchombo, who was at the San Diego University Linguistics Department and later moved to Berkeley to start Chichewa language classes for graduate students and others interested in linguistics. Joan and I learned Chichewa, intending to include it in a comparative study and research on African languages. Sam was our teacher. We embarked on intensive research in these three languages and published our findings in prestigious linguistics journals. We also co-wrote a grant proposal to the NSF, which would allow more time and resources for the kind of research we were interested in.

Joan and I traveled to Tanzania for a month to collect materials on Chaga. We interviewed elders whom we had determined spoke the purest form of the language without mixing it with Swahili. What I also gained from this experience was the history of my community, which I still have on tape for future research projects.

Building the Swahili language courses at Stanford was enjoyable and fun. I had motivated students, many of them from the Linguistics Department. We had several cultural events that allowed students to learn and be immersed in culture, prepare Swahili foods, and create culturally based skits that enhanced their learning. Many of these students were among the few selected to participate in a government-funded, four-week intensive language and culture program in East Africa (Kenya and Tanzania), popularly known as Group Projects Abroad (GPA). Others applied for government fellowships that allowed them to spend time in East Africa doing research for their master's or doctorate degrees in any applicable subject (linguistics, but also history, anthropology, sociology, religion, education, and political science).

The collaborative nature of the Joint African Studies Center for Stanford and Berkeley was interesting but not clearly defined to me at the beginning. I did not realize until later that one of my responsibilities was to serve the Berkeley University African Studies programs. This required me to travel across the bay to Berkeley twice a week to teach a two-hour Swahili class. The logistics were complicated as I had to drive to Fremont and catch a train to Berkeley to avoid the 101 freeway. Furthermore, it was not clear how my EFT was calculated or impacted by assuming teaching responsibilities at both Stanford and Berkeley. My annual salary at Stanford University, after taxes and benefits, was $11,000.

At Stanford, I was teaching eight courses a week (fifty minutes each). With the two Berkeley classes, there was an additional two hours, bringing my weekly teaching assignment to a little under nine hours of teaching, with preparation, grading, and travel time to Berkeley not included. It took me at least two hours to travel to Berkeley and two hours back. I never understood whether this assignment was an overload or just an unpaid requirement.

Regardless, I enjoyed teaching at Berkeley. The students were extremely motivated and fun to interact with. But because I was on campus only twice a week for two hours each, I never connected with anyone on campus. My classes were in the afternoon when most of the Linguistics or African Studies faculty were done with teaching and gone for the day. This lack of connection made me feel like an outsider.

Compared to Stanford, Berkeley was very politically active. There were posters all over campus and on any spare wall, something you did not see at Stanford. I was also intrigued by the small groups of students gathered at different places on campus intently listening to someone preaching. The subject matter varied with some political and others religious. The only time I saw students in a crowd at Stanford was when classes were changing. Everyone was always in a hurry to get somewhere. The clothing worn at these two schools also was different. Stanford faculty members dressed conservatively while the faculty at Berkeley appeared too casual at times. There were a couple of faculty members at Stanford who dressed casually (in jeans and no ties or blazers), but the majority always wore business attire. Often, individuals would comment about those they thought were underdressed and labeled them "rebels." I suppose the First Amendment shielded these rebels from being reprimanded by those who sought to preserve what they considered the "Stanford look."

Despite the different political and dress cultures at each school, what I enjoyed most was the train ride to Berkeley. It gave me time to read, reflect, or simply daydream. I knew that whenever I was on the train, I did not have to make the trip the following day and that break was welcome. At the end of the first year, Dr. Huntington, the head of Linguistics, learned for the first time from the office manager that I was teaching at Berkeley twice a week. She left a message for me to see her at my convenience. I made an appointment to see her a few days later. The first question she asked was, "Is Berkeley paying you for teaching there?" At first, I was not sure why she asked the question with such seriousness. I thought maybe I should have reported the Berkeley service to her. I sheepishly responded, "No."

"Do you have a signed agreement?" she asked. "No," I replied.

"So, how did you end up there?"

"I was informed by the African Studies director that it was part of my assignment as a service to the Joint Center," I replied.

I could see the concern in her eyes. After a moment, she said, "You should be paid extra for that, and you should get mileage for your weekly trips up there. I will talk to the African Studies director about that," she noted.

"Thank you," I said and left.

When I stopped at the office, my friend Gina asked me how the meeting went. I told her what Dr. Huntington had said she was going to do.

"Uhm," Gina grunted. "It is about time. You cannot do two jobs for only one pay."

I do not know when Dr. Huntington spoke with the African Studies director, but I was pleasantly surprised to receive a reimbursement check for the travel mileage and year of teaching at Berkeley. Shortly after, I received a letter from the Joint Center informing me that the following academic year I did not have to travel to Berkeley because they were hiring someone to take over my teaching. On one hand, I missed the two-hour train ride that gave me an odd feeling of calm. On the other, I knew I could maximize my research and teaching time at Stanford. I was happy in both cases. As it turned out, my third year at Stanford was the most productive, research wise.

An Amazing Tour of Three African Countries

In 1985, I met Dr. Musimbi, a linguist, at a linguistics conference organized by Boston University. At the time she was working for United Bible Society, an interdenominational organization that translated the Bible in different languages. The society had regional offices in Africa, and Musimbi worked at one of the African regional offices in Nairobi, Kenya.

After our initial meeting at the conference, Musimbi and I became friends and stayed in touch through emails. We did not see each other again until 1986 when she sent me an email indicating that she was going to be in San Francisco and wanted to come to Stanford for a visit. I was delighted. She was interested in my writings in linguistics, particularly my focus on language gender and culture as well as discourse analysis.

When she visited, she stayed with me for a couple of days and visited my classes and community after-school programs that I was involved in. Over dinner the night before she left, she asked me if I would consider working for the United Bible Society. She told me she had already suggested my name to the director at one of the USA regional offices in New York, and someone was going to call me. She encouraged me to consider the opportunity and, if possible, go through the interview and orientation process. The interview and orientation entailed a month-long tour of different countries in Africa where the society had translation offices. She talked passionately about her work as a linguist in the organization and convinced me that I would be a good fit because of my interest in text analysis. In the end, I relented and promised her that I would consider the invitation if it came, but I would not commit until after the interview and orientation tour. I needed to be fully informed to evaluate my experience at the different centers.

It took a few months before I heard from UBS; 1986 ended and I forgot about the whole discussion. Then in February 1987, I received a letter from the UBS office in Delaware. The letter indicated that my friend had recommended me to UBS and the organization wanted to explore the possibilities. They outlined what the orientation would entail, including a monthlong trip to cover Kenya, Cote d'Ivoire (formerly Ivory Coast), and Ghana (formerly Gold Coast). The goal was to familiarize myself with some of their field translation projects. The letter indicated that all the travel expenses and accommodations would be covered by UBS.

I kept the letter for three days before I decided to respond. During the three days, I searched my soul to see if this was what I wanted to do and whether I would trade Stanford for UBS. I was not sure I wanted to leave the teaching profession. I recalled how hard it was for me to leave my first job with the Ministry of Public Services in Tanzania after completing my second degree, a master's in linguistics, and accepting a research position at the University of Dar es Salaam. I missed the classroom and was not too comfortable exclusively doing research. Nevertheless, there was a part of me that wanted to explore this opportunity before I said no. On the fourth day, I typed out a response letter accepting the offer to visit the centers and to explore the task at hand. I did not mention my conditions for accepting the job. My response must have generated enthusiastic prospects at the headquarters because a flurry of letters followed in a matter of days from different sources—the operations office in Delaware, the foundation's headquarters in New York, and the Africa center in Nairobi. Each letter focused on a specific aspect of the orientation trip. The officials allowed me to pick when I wanted to travel. I chose August 15 through September 15, 1988.

Preparations for the trip included procuring visas for the three countries, a plane ticket, and packing a bag for a month's travel. When the time came, I flew out of San Francisco, and the first stop was Nairobi, Kenya. This was a familiar environment; I had visited Kenya twice before. Kenya UBS arranged pickup from the airport and a hotel stay for the five days I was going to be in the country. I had arrived in early evening and the ride to the hotel was short but marred by traffic jams. Nairobi looked busy, like a metropolitan city with many international offices for companies, agencies, and organizations. There were a few skyscrapers, and the hotels appeared grandiose with at least five to ten floors. The hotel reserved for me was only two floors, with the first floor used for business services and the second for guest rooms. The layout was inviting and very pleasant for travelers who wanted peace and quiet. I met some other international visitors there. Most of them were from England and worked for the British volunteer service, the equivalent of the US Peace Corps.

Once checked in, I went to my room took a warm shower, and then headed downstairs for dinner. What happened at dinner took me by surprise. The restaurant was sparsely occupied with two servers in attendance. I chose a table and waited. The servers were visible from where I was seated, and I could see them looking at me from time to time. As other customers came in, they rushed over, showed them a place to sit, and started serving them. Twenty minutes later I was still not waited on while those just coming in had been served. At this point I was getting concerned, so I raised my hand, and the waiter came to my table. He said hello and waited for me to speak. I decided to take charge and told him I wanted to speak to the manager. It did not take long before the manager showed up. He asked me how he could help me, and I waited for a few seconds before responding. I wanted to make sure I did not explode in the restaurant and cause a scene. I asked him about the hotel rules for service. Instead of using Swahili, I decided to speak English to conceal my identity. He wanted to know why I was asking such a question. I told him I had just arrived from the United States, and was staying at the hotel, and I was curious why I had to wait for twenty minutes without service. I asked him if the hotel practiced racial or gender discrimination. I pointed out the guests at the other tables who came after me had already been served. He apologized and called the waiter back to the table. He asked him why I had not been served and, without hesitation, said he thought I was waiting for someone else to join me.

I did not give the manager time to speak but asked the waiter why he thought so. He said he thought I was from the area and most women who come to the restaurant come with someone. I got the sense that he assumed that I could not pay for my own dinner, and that I must have someone joining me to pick up the tab. In other words, I was not the type of person who would be independent enough to stay in a hotel let alone buy my own meal. I must have been waiting for a man who had the power of the pocket. The manager apologized and decided to serve me himself. He immediately asked me what I wanted to drink and eat. He left to get my order organized. While I was waiting, I had time to review the cultural implications of the experience.

At that time, very few women went out alone, especially in places where men go to socialize. Women, wives, girlfriends, or otherwise are usually accompanied by men who pay for everything. There were very few local women who had prime positions in the government or companies who would be traveling and staying in hotels. Even then, most would be accompanied by an assistant or a family member and, therefore, never alone. This was the same for international companies, agencies, and corporations. Rather than women, they sent men out on trips to represent them. I felt sorry for the young man because he did not know any better. My hope was he would learn from this incident and never again judge a book by its cover. He was a victim of the unfortunate social and cultural norms he had learned from his superiors.

Shortly after my reflections, the manager returned with my drink and then another server brought me my meal. On the next day, when I returned from my orientation visits, I found a basket of fresh fruit and chocolate in my room with a note from the manager: "Complements of the Jacaranda Hotel Management." The note added that my dinner was complementary and would not appear on my bill. I was not paying the bill, but I was sure UBS would appreciate the gesture from the hotel, though they did not hear of the incident that resulted in this free meal. As for the hotel, I thought their gesture was classy and I made sure I expressed my appreciation to the manager. It was all smiles after that, and bitterness forgotten.

I did not see the original two servers again that first evening, but I saw them on other days. Interestingly, the same server was always there in the evening to attend to me. Throughout my stay at this hotel, we became friends and talked in depth about my American experiences. He said he had been accepted at the American University in Nairobi and planned to come to the US for his master's and doctorate degrees when he graduated. Every evening thereafter, I gave him a big tip whenever he served me.

My five days for the Nairobi orientation were interesting. On the first day, I met several UBS staff members. Because this was the Africa Center headquarters, most of the people I met were high-level officials, a mixture of Africans from all over the continent and some Americans. It felt like an intense job interview. The questioning occurred at the different offices I visited and continued through lunch at a restaurant in town. By the time I returned to the hotel, I was exhausted and happy to have quiet time to myself. There were no TVs in the rooms at the time, but the hotel had a small radio on the side table. I enjoyed the African music and news in Swahili, which made me feel completely at home. The basket of fruit and chocolate was an added consolation.

The second day was lighter as we made trips to different sites where the translations took place. There were about eight offices occupied by one translator. There was another large room where new translators were being trained to handle different languages in Kenya and Tanzania. I was told that there were other training places in Rwanda, Zaire (now the Democratic Republic of the Congo), Malawi, and South Africa. My host explained that if I joined UBS and was stationed at their headquarters, I would be sent to the different translation centers to participate in the training and supervision of language proficiency testing of the local translators. In addition, I would work with the local translation supervisor in the certification process of the final draft of the translated Bible for publication. In the case of Swahili, I would play a major role in the final decision for publication since I spoke the language. There was a native speaker for each of the African languages at the headquarters who had similar responsibilities. My best meeting was with these experts because I was able to learn about the challenges as well as the trials and tribulations of overseeing the tasks from the initial translation to the publication of the finished product.

Day three was even lighter. I spent some time in the morning with the translators, and in the afternoon, I was given a city tour. Later that evening, I was invited to dinner at the house of the head of the Swahili Bible translation section. Because he was from Tanzania and a linguist by training, we spent most of the evening talking about our shared connections. Day four was my last full day in Kenya, and I spent more time with the translators and trainers. On day five, I used my time to get ready to depart for Abidjan, Cote d'Ivoire. The flight was making two stops, one in Brazzaville, the capital of the Republic of Congo, then Accra in Ghana before landing in Abidjan.

I was glad when I finally got off the plane. I had flown from the east coast to the west coast of Africa, a distance I took for granted only to find out that I had traveled almost three-thousand miles. Nevertheless, I was ready for the adventure because this was my first trip to western Africa and to a French-speaking country. I was not sure how this was going to go since I did not speak French. Earlier at Stanford, I had signed up to audit a French class, but I dropped it after a week because I did not like the teaching method of mostly drills and memorization. Now I wished I had stayed in the class for at least a semester.

As I exited immigration, I saw a tall, young man holding a big sign with my name on it. I raised my hand, and he came running to greet me and gather my luggage. He introduced himself as Joachim, a native of Cote d'Ivoire and a UBS staff member. He was one of my hosts during my visit to the country. Joachim oversaw the office in Abidjan, supervising about ten staff members who translated the Bible into different Ivorian languages, primarily Mande, Bambara, Daloa, and Betie. He worked in collaboration with the directors in Bouake, another city we were to visit in a couple of days, as well as Daloa, the last of the centers I was to visit.

Joachim drove me to the hotel, called Tjama, and before he left, he gave me a schedule that was similar to the one in Nairobi. He wished me a restful evening and advised me of the nine o'clock pickup time. The hotel was a nice building with a French influence. I had seen that in the city as we drove in and it was even more pronounced at dinner in the restaurant—menu, French wine, French bread, and even the water they served, *Avion*, came from France. I was grateful the menu was in English *and* French. I ordered lamb on couscous, French-style green beans, and a glass of red wine. I took my time with dinner and went to bed soon after.

When I woke up in the morning, I panicked. The amount of sunlight filtering into my room and the noise on the street below from vendors setting up their businesses made me think I had overslept. I checked my watch, but it was only six in the morning. Unlike Nairobi where mornings reminded me of England, overcast and slightly cool, six in the morning in Abidjan seemed like ten in the morning everywhere else. I had planned to be ready and finished with breakfast by eight-thirty. Breakfast was pleasant, with an assortment of breads and rolls, croissants, tropical juices, sausage, ham, and eggs to order. I was not used to this kind of luxury. The experience was making a grand impression on me, but I knew this did not reflect what it would be like if I took the job with UBS. I decided to enjoy the luxury while it lasted.

As I was finishing breakfast, Joachim showed up. We still had time, so I asked him to have a cup of tea or coffee while we talked. On the way to the office, Joachim pointed out points of interest in the city, mostly government buildings that were lined up along the main road. At the office, he introduced me to a staff of five people. He supervised their translations and edited the final versions before sending them to Delaware for publication.

On my first day in Abidjan, I visited with the staff who spoke excellent English. The setup here was less bureaucratic compared to Nairobi. There were more casual discussions between translators about the specifics of their projects. Joachim walked me through their day-to-day activities and the process from data collection to processing and editing the final copies of the Bible. At lunchtime, Joachim and his assistant took me to eat at a local restaurant. I realized that couscous was a popular dish, but there were different varieties. At this restaurant, their couscous was made from cassava (known as "yuca" in America). I was also treated to rabbit for the first time. The cassava couscous tasted different from the wheat-based, which I had had at dinner in the hotel. I liked it so much that everywhere I went I ordered cassava couscous.

After lunch, I went back to take a break. Joachim had invited me to his house for dinner and the opportunity to meet his family. Instead of going to take a nap, I decided to explore downtown Abidjan. I stayed on the main highway into town, which was lined with big shops, bakeries, and government offices. As noted earlier, everything was heavily influenced by the French, but most of the Ivorians wore traditional clothing. The only exception could be found in offices where employees wore western business attire. I limited my tour to half an hour, and then returned to my hotel room and settled down to read a book. At six o'clock, Joachim came to pick me up. On the way, I noticed long lines in front of what I thought was a shop, but Joachim told me it was a bakery. People stopped on their way home to buy fresh bread. I saw some with a long French loaf under their arm, breaking off chunks of it to eat as they walked home. Joachim noted that it was common for people to buy two loaves, one to eat as they walked home, a way to reward themselves for a day's work, and the second loaf to take home to their families. I thought it was an interesting concept.

The trip to Joachim's house took about twenty minutes from the hotel. It was a modest two-bedroom home with a large, well-furnished living room and a spacious kitchen that had a door leading to their vegetable garden. Like many African households, there were two servants, one who helped the head of the household with cooking, cleaning, and taking care of the children when they went to work. The second servant, usually a man, was the gardener who maintained the compound and helped with laundry and ironing. When the car approached the gate, the male servant opened it to let us in. As soon as we were parked, Joachim's two kids, a boy and a girl who were playing in the yard, stopped and ran up to welcome him. They grabbed his arm and gave me a "who are you?" look. Joachim introduced me and told them to say hello. They did so, in English, a clear indication that they were learning the language at school. Joachim noted that they were trilingual and spoke English, French, and Mande, their native language. Both he and his wife tried to talk to them in Mande at home to make sure they retained their roots. They used French in school but learned English as a second language. It was not required in government schools, but his children went to a private school.

As I entered their living room, Joachim's wife, a tall and elegant woman, emerged from the kitchen to welcome me. Joachim introduced her as Rosina. She extended both of her arms for a hug accompanied by double kisses on each cheek. Rosina showed me to a seat and sat down for a brief chat before she returned to the kitchen to continue dinner preparations. The children warmed up to me and eventually sat with me for a while. They spoke more French than English to each other and occasionally said something to me in French. Once they realized I did not speak French, they lost interest and went to play elsewhere. Joachim, who had left us for a moment, came back and invited me to the dining table, which was behind a short wall that separated the dining room from the living room.

The spread on the table was impressive. Rosina was an excellent cook and it looked like a feast with an assortment of African dishes with fish, chicken, and goat meat. They talked about the origins of each dish. The goat meat in thick pepper sauce was from Burkina Faso, the chicken in sauce was a specialty from Senegal, and the crab on a bed of vegetables was their favorite dish. These were served with steamed rice. They also had French wine, tropical fruit salad with yogurt, and tea.

The second day in Abidjan was also light. We started with a morning visit to the UBS office at ten o'clock, participated in a couple of informational meetings, and then did a bit of sightseeing followed by lunch at a restaurant near the beach. Afterward, we walked on the beach and then drove to the cloth market where I bought fabric. We also visited a music shop where I bought Ivorian music cassettes. In the evening, we went to Hotel-Ivoire for an African music concert. I had an opportunity after the concert to tour the hotel complex that resembled the Beverly Center in Los Angeles, California.

It was here that I started questioning the definition of poverty. I could not believe there was an African country with buildings like this but was on the list of poor countries. The people were poor, but their leaders did not show signs of poverty. We left the center and Joachim dropped me off at the hotel. The next day I was scheduled to fly to Bouake, another major city in Cote d'Ivoire where UBS had another translation center. I had done most of my packing with only a few items to put together in the morning.

Joachim arrived at the hotel at six in the morning and my flight to Bouake was leaving at eight. I had plenty of time to get to the airport and check in. Joachim waited until I left the counter and watched me walk to the gate, waving vigorously. I wished I had more time in Abidjan, but I reminded myself that this was not a vacation. I had had memorable experiences here, and everyone I met went beyond the call of duty in their hospitality.

The flight from Abidjan to Bouake was an hour. From my itinerary, I was going to be met by Dr. John Ellington, an American who was working for UBS as a consultant to oversee the translation projects in Bouake. It was raining hard when we arrived. At the arrival gate, I did not see anyone holding a sign with my name. I thought they were confident they would easily spot me since this was a local flight with just a few people landing at a small airport. People were leaving, but no one was trying to identify me. I was still in the terminal standing by my sizable suitcase. A gentleman in uniform with a logo that read "Customs" approached me. He spoke in French and when I did not respond, he switched to English. He must have realized that I was a visitor. He asked me where I was coming from and what was in the suitcase. I was a little puzzled since this was a local flight, but I did not know the laws of the country. Cote d'Ivoire was just as culturally and linguistically foreign to me as England and America. At least in England and America I could speak their language, but in Cote d'Ivoire, I was like a fish out of water. I decided to give an answer that would be more convincing and said that I had arrived in Tanzania first but that I live in America. I added that I had personal items in the suitcase, and I was visiting Cote d'Ivoire at the invitation of the United Bible Society. He looked like he wanted to ask me to open my suitcase when Dr. John Ellington showed up sprinting and repeatedly apologized for being late. He addressed me as Dr. Moshi, a title that startled the customs officer who decided to leave before John exchanged greetings with him. The officer's sudden departure surprised me too and after a welcome hug from John I decided to make a joke about the incident. I said, "Thank you for coming to my rescue John. I was about to be arrested for traveling with a large suitcase."

He looked at me, and then turned around to see where the officer was. Noticing that he was out of sight, he said, "Do not mind them, they always pick on people like you. If you so much as trip, they give you a fine, which you must pay in cash. It is terrible, but it is what it is. I am sorry I was late."

I assured him I was fine, and I was glad that it did not go that far with the officer because he would have been the loser. I have traveled a bit on the continent, and I know how to handle people like him. The best method is to ask to speak to their supervisor, and since they do not know who you are, they never return with a supervisor. John had parked his car near the covered walkway outside the exit door. He handled my big suitcase and dashed into the car because it was still raining. We drove in the rain for about fifteen minutes when we came to a drier area. This part of the country looked very much like Tanzania and the airport reminded me of the Kilimanjaro airport in my home region. The land was endowed with rich soil, lush vegetation, an assortment of tropical fruit trees, and all kinds of small farms growing bananas, coffee, rice, cassava, and sweet potatoes. I told John it looked like I was driving through a familiar place in Tanzania. He laughed and promised to visit Tanzania before he was too old to travel. We both laughed. After the short ride from the airport, we arrived at John's house where we were met by his wife, Joanne, I was going to spend the next two nights with them before returning to Abidjan and then heading to Ghana, the last destination of my trip.

John and Joanne's house was huge by African standards at the time. My room had its own bathroom. It was a welcome change from my stay in hotels over the past weeks. After freshening up, I came out to the living room where John was waiting. His wife served tea and scones and stayed to chat for a few minutes. After tea, John showed me some books, pamphlets, and other UBS literature. He invited me to see his office, which was a designated part of the house. He had an assistant who occupied the front, and his main office was in the back. It was a large room with lots of shelves that held stacks of files. He explained that these were drafts of different stages of the translated Bibles. John led me back to the main part of the house where he left me to look through the materials he had shared, and he returned to his office to finish some work before lunch.

Joanne made lunch, which was couscous with chicken, sauce, fresh vegetables, and fresh fruit for dessert. After lunch, John went back to his office while Joanne and I drove to town to sightsee. Bouake was significantly smaller than Abidjan. The town had a few good stores, but some parts were populated with lower-income shops and homes. The more affluent people, like John and his wife, lived in the suburbs with big houses on spacious lots. I was impressed by the town's infrastructure. In comparison, Tanzania was emerging from a depression and the roads and buildings were in terrible shape. The situation in Tanzania was dire and many cities were experiencing shortages of food and necessities. The smaller villages were better off because they grew most of what they consumed, and they could do without a lot of what those who lived in cities and towns considered essential.

After walking around Bouake for an hour, we stopped at a bakery and Joanne bought some bread, confirming what I had seen in Abidjan. Late afternoon and early evening were when the bakeries had freshly baked bread ready. No one bought bread that was not piping hot from the oven, which certainly lived up to the definition of "fresh bread," Joanne explained. On the way home, Joanne asked me about my family. She told me she and John had two grown children, a son, Mark in the United State, a Marine who lived in Miami, and a daughter, Beth, who worked in Tampa, Florida. They also had adopted two children, Joseph who was fifteen years old and Vietnamese, and Rebecca who was eleven years old and Korean. They attended an American boarding school in town, but Joanne and John had arranged to allow the children to live at home, not on campus. I saw the kids in the evening, and they talked about their school and classes. Both were straight-A students, to the pride of their parents. After dinner, we watched a film from their video collection, *My Fair Lady* (1964). I had not seen it before, so it was a treat.

The next day started at nine o'clock with a simple breakfast of homemade muffins, tea, and a banana. Finally, things were close to normal: No more lavish hotel breakfasts like in Abidjan. After breakfast, John and I went to town to confirm my return ticket to Abidjan and connecting flight to Accra, Ghana. We came back and worked in his office until lunchtime. We discussed a few things concerning translation, specifically the areas in which he had worked. He described his work with translators and how his supervisory role covered a spectrum of countries: Mali, Senegal, Guinea, Burkina Faso, and Liberia. He also was the regional administrator, so he filed forms, planned travel, balanced budgets, taught translators how to use computers and what software was necessary, taught interpersonal relations, dealt with common problems in translations, and oversaw the final stages of the translations before publication. He walked me through UBS' responsibilities, which included helping the translators who were not fully supported by the churches that had requested the translation work. The support these translators received was in the form of both guidance and money.

Listening to John present this mini lecture made me consider the workload and responsibilities that rested on his shoulders. He asked me what I thought about this line of work and added that he had requested additional staff and, if I accepted the job, he would request that I be posted to Cote d'Ivoire. I commended him for all his work and carefully avoided commenting on his plan. I was not sure I would be able to do all this work in a foreign country. John was well traveled, a retired Army officer, and more used to working under pressure.

After lunch, we took a break until around three, and then went back to his office and checked translation manuscripts on the computer using Concordance, a lexicon of all the words used in the Bible. We worked until dinner time. I was so exhausted that I felt my head would explode if I looked at any more text. After such a hard day's work, Joanne had prepared a rewarding dinner. She said it was a Senegalese specialty made of chicken marinated in lemon, garlic, and onions. She used a variety of African spices, including turmeric, and thickened the sauce with coconut milk. She made steamed rice and cabbage, and for salad, there were raw carrots. For dessert, she served tropical fruit salad.

After dinner, Joanne and her daughter went to an open house at school. John insisted I finish watching the film I started the night before. I know he was trying to keep me entertained but I knew my weakness while watching movies. If it is not an action film or super funny, I fall asleep in no time. I had embarrassed myself the night before by falling asleep, and I did not want to repeat it. But I had to respect my host, so I agreed. I prayed that I would stay awake after such a nice but heavy meal. Miraculously, I managed to sit through it, but to tell the truth, I found the film boring.

The following day was Sunday. One would think that because John had an office in his house that he would not drive to the headquarters in town, but he did. I was surprised at his discipline; this was a day for his family. It was also his birthday and Joanne threw a dinner party for him. Quite a few of his colleagues came and brought food and presents. This was my last night with John and his family. I was about to visit another translation project, and John was to drive me there in the morning.

We left their house at 9:30 a.m. to drive west from Bouake to Daloa the city on the west coast. Driving there, we passed several police checkpoints. John explained that there had been a coup in Burkina Faso where the army overthrew the government and deposed President Thomas Sankara in 1987, the year before my visit. He said that Cote d'Ivoire did not want bad people from Burkina Faso sneaking in across the border. Despite the police checkpoints, I liked the countryside and the small communities along the way. We arrived in Yamoussoukro where the president of Cote d'Ivoire, Félix Houphouët-Boigny, had built a Catholic Basilica, called the Basilica of Our Lady of Peace, and his presidential palace. John suggested that we stop first at the basilica and then the palace. This basilica was an amazing and imposing structure for Yamoussoukro, a small city compared to Abidjan. The grandeur of its architectural design surpassed Saint Peter's Basilica in Rome. The president of Cote d'Ivoire was Catholic, and it was believed at the time that he built this cathedral to impress Pope John Paul II, who also was invited to the inauguration ceremony. Rumor has it that the pope was unimpressed because he could not condone the Catholic president's use of government funds that were earmarked for the welfare and development of the Ivorian people to feed his ego. John took me inside the basilica to see the wonders within. I was awestruck by how exquisite it was. I was used to the modest buildings we called cathedrals, churches, or chapels. A basilica was a new level in my exposure to the Catholic Church. The stained glass, the altar, and the way the pews were arranged in the massive space were breathtaking. This building could easily hold eight-hundred people. We exited the basilica and walked around to see the architectural design. It stood on a large plot of land that was slightly elevated, giving the basilica a view of the main road. The back offered a panoramic view of the land below. It was impressive.

We left the basilica and continued to the presidential palace. John explained that the president built this as a vacation home that would become his permanent residence when he retired. It was a large piece of property, and the buildings befitted the palace title. I was interested in a large pool-like structure in the middle of the compound with chairs around it. It looked like a swimming pool, but it was dry. I saw crocodiles at the bottom of the pool. Before I could ask, John told me how the pool worked. It was a few minutes before noon, the time when the crocodiles were fed. John told me that when the president was at the residence, he came to the pool to see the crocodiles eat, which explained the chairs. The crocodiles were fed live chickens and the president took pleasure in watching the crocodiles chase them while the audience cheered. John noted that if the president was not there, the crocodiles were fed different kinds of meats. By this time, I was feeling sick to my stomach. John noticed my discomfort and suggested that we leave. I was grateful. I did not say much until we got back in the car. I was thinking about the many chickens being fed to the crocodiles by the president for sport. I was sure there were some people who could not afford a chicken. I remembered that we had chickens at my parent's house, and we only killed one to eat on special occasions or when we had unexpected, distinguished visitors and there was nothing in the house that my mother could serve. Once in the car, John asked me if I was feeling better and I assured him I was, and we continued our trip to the next destination.

John drove to Daloa, the third and last translation center for my visit to Cote d'Ivoire. This center was under the supervision of Lynell Georgie. Like John, she had a big house with a section designated as her home office. They also had reserved a room for me at their house. Lynell was an American married to an Ivorian, and she was the director of the translation project at Daloa She was expecting us for lunch and had prepared spaghetti, salad, and French bread. After lunch, John left for Bouake and Lynell took me to her office to meet the two translators who worked under her supervision and who were sponsored by a church in Liberia. Lynell wanted me to take it easy until around four o'clock when her husband returned from work. They wanted to take me to her husband's family home to meet his extended family.

Their family home was a semicircle of small buildings in one compound owned by the male sons of the family. The sons lived there with their wives and children. There was one house in the middle, which I was told was occupied by their father, the head of the family. This demonstrated the role of the head of the household and the culture in general. This patriarchal unit included seventeen wives and over forty children. The father lived for many years, dying in 1986, two years before my visit to Cote d'Ivoire. Culturally, the female children were not valued the same as the males and did not receive any family inheritance. When they married, they assumed a place in their spouse's household by virtue of their male offspring.

I had a chance to meet two of the sisters who lived in the area. One of them made lunch in my honor one afternoon and sent it to Lynell's house. Lynell and her husband did not live in the family compound, but they owned one of the units that they used when they were home with the family. It was interesting to see how the strong and tightly woven cultural fabric of this family in the compound was displayed. I wondered how Lynell, an American, adapted to this culture, which was so different from her own, and one that embraced American exceptionalism at that. Lynell had one son, Karibouh, or George II. His dual culture was apparent in his names, one traditional and one western. I wondered whether he would adhere more to his father's culture or gravitate toward western culture, as many young people of his generation did. It was also possible that he would be the exception and, instead, proudly embrace both.

Tuesday was my full day of work in Daloa. Like John, Lynell had prepared some translations I could work on with the staff. We started at eight-thirty; the translators wanted to work on the Gospel of Mark. I was amazed at their experience. In two hours, we were able to translate three chapters. We took a break and were served tea and cookies that Lynell's housekeeper brought to the breakroom. We resumed at eleven o'clock and worked until lunchtime, which was around one o'clock. We were finished to chapter five and did not resume until three in the afternoon. Then, we worked until five-thirty, with another tea break at four o'clock. By the time we finished for the day, we had translated seven chapters. I was impressed at how hard the translators worked. I learned later that they had deadlines for each assignment that they tried to meet, if not beat. I wanted to know how much they were paid per month for what they did. I was not surprised that as Africans they were paid the local minimum wage, less than a dollar a day, but as an expatriate, the salary was based on the market value in the United States. My two hosts from the United States, however, John and Lynell, received a wage based on what they would have earned in the US. In addition, they had a daily per diem to cover living expenses: house, food, utilities, transportation, and medical. I wondered whether I would have been considered an expatriate or a local staff member if I accepted a job with UBS since I was not American.

After the translations, Lynell thought we should go for a walk since we had been sitting all day. I welcomed the idea. We returned to the house to get the car and go grocery shopping for dinner. On our way back, we stopped at her sister-in-law's house for a brief visit, and then headed home to cook. We had a simple dinner that night and then retired.

The next day, after breakfast, we started work again at eight-thirty with an identical schedule to the day before. To my surprise, we cleared seven more chapters. All the completed work was given to Lynell to check and then she had a group meeting to go over the changes she made. She showed me how she checked the translations and what she considered critical changes to ensure coherence. The tasks were labor intensive from start to finish with pressure to complete the texts to meet the publication schedule. This was my last day at Daloa before I returned to Bouake to spend a night at John's house before flying, once again, to Abidjan and then Accra, Ghana.

On my departure date, we left late morning, and Lynell's husband drove me back to Bouake. We only made one stop for lunch at his aunt's restaurant and arrived at John's house in the afternoon. Lynell's husband did not stay long because he wanted to get back to Daloa before dark. John and I went to the bank so that I could exchange my money. Unfortunately, I was unsuccessful because the bank was not accepting American dollars. Apparently, there had been an incident where someone had exchanged some fake dollars and the management had closed the foreign currency section until further notice. Since I only needed a small amount for incidentals on transit, John agreed to take my dollars and gave me the equivalent in local currency. That evening I was surprised to learn that John's wife had planned a farewell dinner for me. I had a chance to see all the staff I had met at his office. It was a good ending for my tour in Cote d'Ivoire.

My flight to Abidjan the next day was mid-morning. John dropped me off at the airport, waited for me to check in, and then waved for the last time. This was a work trip, but I felt sad as I waved good-bye. On my way to Abidjan, I started evaluating my experiences in Cote d'Ivoire. It was the most productive so far. I learned a lot by doing and experiencing the culture, particularly from the home stays rather than hotel stays. All the people I met in Cote d'Ivoire were gracious, accommodating, and extremely hospitable. What surprised me was how the Americans I met and worked with had assimilated into African culture. Their assimilation combined with their own goodness made my visit memorable, both in learning from and having a good time with them.

My hosts were generous and helped me to learn a lot about Cote d'Ivoire. I used what I saw and experienced to evaluate life in Africa, particularly in Tanzania.

There were some similarities between Tanzania and Cote d'Ivoire that informed my understanding of what happened in Tanzania between the 1960s and the '70s when the Chinese and Russian influences were the strongest. This was an opportunity for the western nations to use African countries as the battleground for the Cold War. The villagization in Cote d'Ivoire, which is the process used by the government to resettle people into designated villages, resembled the *ujamaa* villagization I experienced in Tanzania when I was growing up. I was old enough to see and remember what happened when people were forced to leave their villages and move to undeveloped areas to start over under the guise of overcrowding. The government forced people to establish new dwellings by building houses and then use assigned land to establish small and large farms for both individual and communal farming activities. With communal farms, villagers spent a certain amount of their time working on the land that was collectively owned with the promise that the crops would be divided equally, regardless of how much effort one put into the production. That was the spirit of *ujamaa* or "familyhood," a culturally embraced concept but one that was colored by aspects learned from China to create a national government policy. The emphasis on the communal farms was to produce commercial crops that were sold to cooperatives run by the government and, theoretically, the revenue was to be divided equally among the participants. The communal farms and the cooperatives failed because those who ran the operations mismanaged the resources and revenue, therefore bankrupting the cooperatives and demoralizing the participants who had sold their goods to these cooperatives but never got paid for their efforts.

There was another, more important side to this: the human costs. During the forced relocation in Tanzania, many women lost their lives, giving birth on the roadside or in makeshift accommodations at their new locations. The government tried to implement the policy without adequate preparations. Ideally, the government should have developed the new villages with housing, a fully realized infrastructure, water, electricity, medical facilities, access roads, and schools rather than assuming that these could be done by the villagers after they had been relocated. Furthermore, when the villagers in Tanzania resisted being forcibly moved, their houses were torched, making it impossible for them to stay. Some villagers returned to their old locales to get whatever produce was left on their farms to sustain themselves in the new location, but their extended absence created opportunities for thieves to shamelessly help themselves to their produce.

Cote d'Ivoire's villagization process had some good ideas compared to Tanzania. While Tanzania moved people to undeveloped areas and expected them to fend for themselves by building new residences, schools, and medical centers, Cote d'Ivoire built the houses before forcibly moving the people to these new locations. However, they did not complete the infrastructure. The necessities like water and power were not put in place, nor were schools, health centers, shops, and markets for bare necessities. The villagers were expected to travel back to their original villages to obtain these goods until they were able to develop the new place. Ethiopia, like Cote d'Ivoire, emulated Tanzania's villagization policy but failed miserably because they made the same mistakes.

I recalled Dr. Ellington talking about what the Ivorians thought about the unoccupied modern houses built by President Houphouet-Boigny's administration. They named them the Houphouet-Boigny's failed villagization experiment. I wondered whether all the countries that tried this failed because of poor planning or the eagerness to emulate China. China experienced a housing problem because of overpopulation. In Tanzania, Ethiopia, and Cote d'Ivoire, the problem was an abundance of land that needed to be populated. The idea was that by thinning its population from the existing villages, they could establish these new areas and solve their underdevelopment problems. In other words, the villagers themselves would be able to develop the new area. This could have been done by creating incentives for people to move, allowing the area to slowly populate rather than forcing people to abandon their developed farms and homesteads to satisfy a half-cooked government policy. History has shown, for example, if good roads or railroads are built, people would willingly move closer to the areas served by them because transportation would be guaranteed, which is the key ingredient to development. Factories or industrial complexes depend on good transportation and people will willingly move to gain employment to improve their livelihoods and those of their families. The lack of foresight and good planning by these governments were the sole reason for the failure of the villagization programs and policies.

These thoughts and comparisons made me think about the African leaders' excessive reliance on foreign examples to solve problems specific to their local conditions. Colonialism seduced many leaders and as such, they continued to lead their citizens without longer-term plans informed by local conditions. These leaders were fueled by greed. They thought they could fly when they had no wings. The colonial legacy robbed Africa of its vibrance and ability to make decisions that were sound for its people, thus allowing dependency to replace creativity and self-worth. I was utterly wrapped up in these thoughts, which made me lose all track of time. I was startled when the pilot announced that we would be landing in Abidjan in five minutes. I did not get off the plane because it was continuing to Accra.

The flight to Accra, Ghana, left on time and in an hour, we were on the ground. As soon as I got out of the terminal, I saw someone holding a sign with my name on it. I waved and the young man came forward to meet me. He did not say anything, just grabbed my big suitcase and led me to the car, a Land Rover that was parked close to the arrival gate. In the front passenger seat sat a man dressed in traditional Ghanaian clothing, a big piece of kente cloth that dropped the length of his trousers and was half-folded on one arm to expose his arm and shoulder. He opened the door and came toward me, extending his hand in welcome. He introduced himself and said he was the UBS representative in Ghana. I thanked him for coming to meet me. He opened the passenger door for me and, once settled, he closed it. The young man who met me was the driver.

There was complete silence during the ride. I started comparing my arrival in Cote d'Ivoire with my new destination. I focused on observing the landscape as we drove through the town to the suburbs. We arrived at a big house with an iron fence around it and a decorative gate. As soon as the car approached, a young man behind the gate opened it and we drove in. The compound was adorned with Ghanaian art and the lawn was green and neatly manicured with scattered rosebushes of various colors. The whole appearance was impressive. I was not sure if this was someone's home or a hotel/motel. My host got out of the car and headed inside the house, entering through a decorative door with Ghanaian Adinkra symbols (symbols that represent concepts or aphorisms; they are used on fabrics, furniture, logos, and pottery, and are incorporated into walls and other architectural features). He did not say a word to me. The driver came to my side of the car, opened the door, and said, "Welcome." I thanked him and proceeded to get out of the car. I did not know what to do, so I decided to stand there and wait for instructions. At this point, I was feeling uncomfortable and wondering if I was going to survive a week in Ghana. The driver took my suitcase inside the house and then came back, summoning me to follow him. I walked in through the magnificent door, admiring the carved symbols. The young man showed me a chair and then disappeared through another door. This looked like a small living room, slightly dark without the lights on since the draperies were drawn keeping the outside light out. I sat there for ten minutes before the side door opened, and another young man came in holding a tray with a glass of water on it. He stood at a distance and just stared at me. I was not sure what to do since I had not asked for water and was not sure if this was a cultural ritual that I was supposed to know about but did not. I looked away and waited. I looked back and the young man was still standing there without a word. When I looked at him again, he inched closer. Common sense was my only guide here, so I stood up, and took the glass of water, and went back to my seat. I sat down holding it in my hand. The young man continued to stand there. I relied on common sense again, and I took only a sip since I was not thirsty, stood up, and placed the glass back on the tray. He bowed his head and left the room. Though I did not realize the gesture at first, I threw my hands into the air both in amazement and frustration at how this trip was going. I started to wonder what was coming next and how I was going to survive it.

Suddenly, the same door opened again, and my host, the man from the Land Rover, came in followed by his wife and children. He had changed from traditional Ghanaian garb and was dressed more casually. He greeted me like he was seeing me for the first time and introduced himself as Kwaw Mensah, his wife, Nana, and children Kofi and Samuel. They all sat down but only his wife spoke, telling me she was happy to have me in their house and that I was going to be there for the duration of my time in Ghana. Shortly after that, her husband and children left, and Nana and I sat for about five minutes talking. Then she took me to the room they had reserved for me. The house was large with a family wing and a guest wing. There were three self-contained guest rooms on this wing, and they had given me the corner room with a window facing the garden. The draperies were opened, and I could see the landscape and the flower beds with roses. My heart was uplifted and the discomfort I was feeling started dissipating, giving me a glimmer of optimism. Before she left, Nana told me I could take as much time as I wanted to freshen up. Dinner would be at eight o'clock, but if I wanted to join her in the kitchen, I was free to do so. I decided to take my time, get over my initial culture shock, and venture out in the remaining daylight. There was a feeling of comfort in this room—the bed, a side table, the lights, and its sheer size met hotel standards. Ghana has a strong British influence and, as such, affluent people live well, combining their cultural assets and a foreign touch with finesse. The bathroom fixtures were exquisite, and they read, "Made in England." The shower and bathtub were separate, a common arrangement in England. There was a hot-water tank on the wall, which needed to be switched on fifteen minutes before hot water was available, and a card on the wall with instructions. The soap dish held a bar of Yardley soap from England, which I was familiar with. I figured this was for the face and hands since there was another bar of soap in the shower that read, "Imperial Leather," also commonly found in England and the British colonies. As Nana suggested, I took my time to wind down and enjoy the room.

    I was sitting on the chair near the window when I heard a knock on the door. Nana came to get me for dinner. The spread was impressive. The dishes were typical Ghanaian: pounded yam, beef cubes in peanut sauce, fried fish sautéed with ground pumpkin seeds, and fried plantains. It was a wonderful welcome dinner. At dinner, Kwaw introduced me to Mr. Roberts. He was the head of the translation. I learned then that I was going to spend most of my orientation time with him.

After dinner, the men left and just Nana and I were left sitting in the living room. She had help, a young girl who cleared the table and washed the dishes. The two boys left for their rooms, but I was not sure where Kwaw and Roberts went. During our conversation, Nana mentioned that there was a wedding the day after tomorrow. She invited me but warned me that no western attire was welcome. She asked me if I had a traditional dress from Tanzania. I had one in the US, but I did not think of taking it on the trip. She suggested that in the morning we would go to the market, buy some material, and take it to a seamstress to make me an acceptable outfit. I agreed.

Friday morning at breakfast, it was just me and Nana. The children had left for school and Kwaw was nowhere to be found. I suspected his wife must have told him we were going shopping, and he suspended any orientation plans he might have made. I did not see Mr. Roberts at breakfast either. The choices for breakfast were very British: cornflakes with hot milk, bread with butter and jam, or scones. That suited me fine. Though the cuisines were different, staying with families in Cote d'Ivoire prepared me for Ghana.

After breakfast, we headed to the market. The family driver drove us, and we spent three hours looking for fabric. Nana picked some out for me that were the wedding colors. We left to visit the tailor she knew and impressed upon her that it was important that I got the outfit by the next morning, the day of the wedding. The tailor promised to do her best and Nana picked a uniquely Ghanaian style she thought would look good on me. She measured me and we left. I was overwhelmed by Nana's generosity. She said that because she had invited me to the wedding and I was not planning to wear a Ghanaian outfit, all expenses were on her. I offered to pay half, but she would not hear it. This was her gift to me from Ghana. I had no choice but to say thank you several times. They talked about different ideas and Nana made additional suggestions. Listening to them talk, I knew I was in for a treat to experience a traditional Ghanaian wedding with some western add-ons. The relative offered a light lunch before we left. By the time we got home, it was three o'clock, and too late to do any official business. On our way home after shopping, we stopped at a relative's house who was involved in organizing the wedding.

Mr. Roberts was at Kwaw and Nana's house when we got back. He had made a schedule for what we were going to do starting Monday and offered to walk me through it and answer any questions I had. His demeanor was different from that of the other directors I had met, particularly in Cote d'Ivoire. We discussed the itinerary for half an hour and then he talked to me about the work UBS did in Ghana. The plan for the week I would be in Ghana was no different from that of Cote d'Ivoire, except that Roberts did not require me to do any translations with the translators. Instead, he wanted me to help with administrative tasks that required filling out forms for different requests and bill payments. With such a flexible schedule, I had time for sightseeing with Nana. We developed a kind of closeness that at times seemed like we had been friends for a long while. We laughed, made jokes, and visited several of her friends who had businesses in the community. Nana had a business of her own, a hair and nail salon, but she did not have to be at her shop all the time. She had an administrator, a business manager, hair stylists, and manicurists. She went to the shop only if she was contracted to style the hair of a wedding party.

The Saturday wedding was a revelation for a first-time visitor from another culture. Many African countries embrace both western and local cultures in their wedding ceremonies and festivities. The western features are evident in the wedding couple's attire and the incorporation of bridesmaids and groomsmen. The church ceremony also had adaptations from western Christian religious practices. In Ghana, a western-style wedding is called a "white wedding," which refers to the wedding dress. A traditional wedding includes expensive clothes, locally made gold and diamond trinkets, beads, and necklaces for the wedding couple, as well as special drinks offered as gifts to the in-laws.

This wedding included practices that were not part of Tanzanian culture. There were Ghanaian rituals performed throughout the week that culminated on the wedding day. I did not attend these because they were reserved for family members and close friends, but I attended the church and the rest of the ceremonies.

On the wedding day, the couple started by going to church and having a clergyman officiate the wedding. They were Catholic, so there was a Catholic service where they exchanged their vows and signed the official marriage certificate. The groomsmen wore black tuxedos with ocean-blue shirts while the groom wore a black tuxedo with a white shirt and a black bowtie. The bride wore a long white dress and a veil. The bridesmaids wore long, sky-blue gowns to match the groomsmen's shirts. The guests, including the parents and relatives from both sides, wore traditional Ghanaian clothing. After the church service, all the guests went to the reception hall while the bride, groom, bridesmaids, groomsmen, and their families left for an hour to take pictures and to change from western to traditional garb.

The waiting ended when we heard loud singing and ululations from outside the reception hall. The master of ceremonies announced the arrival of the newlyweds. Two photographers dashed into the hall, and once they cleared the entrance, turned around and began snapping pictures and taking videos. The groomsmen came in first, dressed in new ocean-blue outfits made from Ghanaian fabric. They came in dancing to the DJ's music while the audience clapped and danced in their seats. Each of the groomsmen had a beautifully wrapped gift that they kept tossing up in the air and catching before it fell to the ground. This act, combined with the dancing, made them look like acrobats. At the end of this line of twelve groomsmen was the groom himself. He too had changed his clothes and wore an elegant Ghanaian ensemble made of kente cloth. The choreography displayed by the groomsmen and the groom was a true statement of how Ghanaians revered their culture. It was ceremonies like this that exposed the richness and warmth of the people and their traditions. The groom looked like royalty in his brightly colored kente cloth and was adorned with gold jewelry and other regalia. Gold is plentiful in Ghana. One can easily find a goldsmith forging ornamental artifacts from the precious ore. The amount of gold worn at this wedding reminded me of why Ghana was named the "Gold Coast" by the colonialists. The name was changed to Ghana after independence. If Ghana's national wealth were in their hands, there would not be a single poor person in the country. But colonialism and inherited policies that favor the rich created a system of the haves and the have-nots, a common phenomenon in many African countries.

The wedding party danced in circles and in front of the in-laws, who were seated on opposite sides of the reception hall. The groom shook hands with his in-laws and waved at his relatives. While dancing, the groomsmen led the groom to the high table designated for him, his bride, and their best man and maid of honor. To respect his bride, the groom did not take a seat but remained standing and continued to clap as the next group entered. Entering the hall in a similar fashion were the bridesmaids who also had changed outfits into long yellow gowns made of Ghanaian fabric. They were holding custom-made fans of yellow feathers, waving them in front of their faces as if attempting to cool themselves in hot weather. At the end of the procession line was the bride who was dressed in the same brightly colored kente cloth that matched her new husband. Likewise, she was adorned with gold jewelry and other regalia. As soon as she entered the doorway, the crowd erupted in a chant and ululation that coincided with the DJ's music selection. The combination of sounds and her small, short dance movements made her appear like royalty. The more the crowd cheered, the more the bridesmaids continued dancing. They danced around the tables where the parents of the groom and bride were seated, and the bride curtsied before each table before she took her seat. Her new husband, who was waiting eagerly, was led by his best man to identify her in the group of bridesmaids. This was part of the ritual, but one of the women sitting next to me said the groom is warned not to take anything for granted. The girl in the same outfit as him could indeed be his wife or they could be playing a trick on him and switch her with someone else. My host told me that there had been a previous wedding ceremony where they wanted to tease the groom and dressed another girl in the expected clothing and let her enter the reception hall. The groom assumed it was his bride and held her hand to lead her to the high table. As they were walking to the table, the decoy removed the fan from her face, turned around to the audience, and started to wave. The crowd let out a loud cry of surprise that forced the husband to look and realized that he had been duped. So, at this wedding, when the groom went to identify his wife, he did not touch her until he had an opportunity to see her face. Then he swept her from the ground and danced to the head table, carrying her all the way. The bridesmaids, joined by the groomsmen, continued to dance, and escorted the wedding couple to their table. They all danced as the audience shrieked and clapped to the music. Once the wedding couple was comfortably seated at their table, the groomsmen and bridesmaids retired to their designated seats. At this point, the music stopped, and the master of ceremonies took over to introduce the family members to the audience.

The introductions signaled the start of the traditional portions of the ceremony, which included gifts of gold, diamonds, beads, money, drinks, and expensive cloth. The gifts were lavish. The guests came up singly and in groups, dancing to the DJ's music and encouraged by the cheers and claps of the audience. Some people stood up and danced in place. I was tempted and encouraged by the host, so I did. It was the wildest party I ever attended. There were two tables in the corner of the hall near the wedding couple where the guests placed their gifts after shaking hands with the newlyweds. Expensive gifts were handed to a designated young man and woman dressed in Ghanaian attire. They looked like a prince and a princess with their half-smiles.

Community leaders gave the couple experienced advice, and they emphasized the importance of family and hierarchies. The advice favored the groom by emphasizing the services and sacrifices required of the bride to make the marriage work. The groom was reminded of his main responsibilities to provide shelter and security for the family and to make sure he was a strong head of the household. One could argue that these tasks were loaded when analyzed, but I was not sure if he thought about his responsibilities this way. One would hope his parents had set good examples and that he was a responsible person by now. Chances are that formal education and worldly ambitions may have interfered with family traditions and values. However, I was not judging this young man but praying that my pessimism was wrong since the newlyweds had strong family support and parents who were revered in the community, judging by the size of the wedding reception, which included at least two-hundred people.

After the community leaders spoke, the wedding couple was escorted by their groomsmen and bridesmaids to approach the parents' table and receive their final blessings before leaving the reception hall. As they walked out, I saw the bride wiping away her tears, a common sight at weddings. Walking out and leaving everyone seated symbolizes the couple's final separation from their old life to their new life. The bridesmaids and groomsmen followed them at a distance clapping and dancing. My host signaled it was time for us to leave. I was glad because it had been a long day and I was ready to go to bed. The following day was Sunday, a good time to take it easy before another workweek.

As expected, Mr. Roberts took me to his office on Monday, and I worked on the administrative tasks he had planned for me. During our tea break, we talked about my plans to join UBS. This was only the second time Mr. Roberts and I sat together and talked for an extended period. He told me how he joined UBS and why he chose that line of work. Originally, I thought he was educated in England, but I found out he had studied in the United States and when he graduated, he wanted to work to gain some experience before returning home. His first job was as a clerk at UBS. At the end of his first year, they asked him if he wanted to join them permanently, which would include working in Ghana. He thought this was a dream come true, and he worked for UBS for the next twenty-five years. The story was encouraging, but I was not sure about the actual work. I had the linguistic skills for what UBS wanted but I was not prepared to assume the responsibilities akin to those I observed on this orientation trip. I was also unsure of where UBS would post me since I would not have a choice in the matter once I accepted the job. I would have worked at the headquarters for a couple of months and then received my new post. Another thing that Mr. Roberts said to me that fueled my hesitance was that his salary was based on the Ghanaian salary scale for his qualifications. Additionally, he received a small housing allowance, which allowed him to have the office as part of his house along with a transportation allowance. Unlike his counterparts, the expatriates, he did not have health insurance because as a Ghanaian he did not need it, and he did not have a paid trip to the United States with his family once a year since he was not a foreigner in Ghana. This reminded me of what I had learned in Cote d'Ivoire, and I was ambivalent about the distinctions based on nationality.

My time on this trip was coming to an end. Before my departure, Nana offered to take me to Elmina Castle, a historic place for Ghana that reminded them of the many Africans who left the continent for the new world as slaves. Slave trade was not part of the colonial education curriculum, so many of us who went to school before independence did not hear much about it. The only time we learned a little about it was in a Swahili language class from a class reader *Uhuru wa Watumwa* or *"Freedom for the Slaves."*[4] The teacher did not discuss slavery as a practice but let us focus on the sufferings of the characters in the story and how they longed for their freedom. Because the book was taught in abstraction, we did not relate to the origin of the characters in the book and focused on the main characters' suffering. The teacher was as removed from the concept of slavery as we were and lacked the passion to discuss what was the biggest atrocity committed by the Arabs and Portuguese to the East and Central African people. In the teacher's defense, teaching slavery was not in the curriculum. Instead, the entire education system was about Europe, its greatness, and its might. The teacher did not say anything about how the continent of Africa facilitated that greatness

---

4   *Uhuru wa Watumwa*, [*Freedom for the Slaves* (1934)], a historical novel written by James Juma Mbotela in Swahili (Kiswahili) and published in 1934 by Sheldom Press for the East African Literature Bureau. Subsequent editions include 1949, 1951, 1956, 1959, 1960, 1967, and 1970. In the novel, Mbotela provides the reader with a chronology of events through the lens of an enslaved person who was captured in Mpanda, located on the interior western part of Tanzania. He traveled through the hinterland to Bagamoyo, on the east coast of Tanzania, one of the entry points for Arabs and Portuguese who traded in ivory and slaves on the east and central coasts of Africa in the late 19th century. The city of Bagamoyo derived its name from a Swahili combo word, *bwaga-moyo*, which has several different interpretations, including "lay down your heart," "cast off heartache," and "abandon all hope," a cry of desperation from the exhausted porters coming to the end of an arduous journey or by the captive slaves losing their hope of ever seeing their ancestral land.

and might enjoyed by Europeans. For the African continent, the emphasis was on poverty, and the need to survive and, hopefully, achieve what Europe had done. The training focused on creating bureaucrats who would continue to accommodate Europe's necessities for survival of its control of the continent's resources.

Elmina Castle in Ghana reminded me that there was so much I still did not know about my ancestral country of Tanzania, let alone the entire continent. I still did not know much about Mpanda and Bagamoyo, the two main places discussed by the author of *Uhuru wa Watumwa* in the book I had read in high school. These two places were just as foreign to me as Elmina, Ghana, Cote d'Ivoire, and many other places. I remembered that for my final examination in high school, I wrote two essays, one on the Ruhr coal fields in Germany and another about cotton production in the United States and its relation to the clothes I was wearing. For the exam, I was able to draw a map of Germany, locating the main towns and cities the train passed as it ferried its coals from the coal fields. To date, I do not know why that was important except to make me appreciate the symbiotic relations between what we could not make and what we consumed from Germany, our former colonizer. The essay I wrote about cotton in the United States became painful as I realized that, at the time, the curriculum did not include that the cotton Tanzania produced was shipped overseas and then imported as clothes at higher prices than what the farmers were paid for their cotton. I was now educating myself on what was left out of the curriculum. The tour guide at Elmina Castle took time to explain each area we visited, connecting the history and the harsh realities.

When we exited Elmina all the visitors looked somber. No one was speaking, and in fact, a few were crying. I walked silently next to Nana, and she did not say anything either. The driver saw us coming and opened the car doors for us. We climbed in, still silent. I decided to break the ice by saying, "Thank you, Nana. That was hard to take but the past is the past, and we need to think of a way to make the future better."

Nana turned back and said, "We hope so, but I do not know. We do not seem to learn from our mistakes. Some of our leaders are still numb to this. The sufferings in Africa have never stopped."

I was tempted to continue the discussion, but I changed my mind and did not say anything else. Nana told the driver to take us to a restaurant owned by a friend of hers. Her friend was not there when we arrived, but the manager received us kindly and served pounded yam, red rice, fried plantains, and fish. After lunch, we drove back home. Because I was leaving the next day, I used the rest of the evening to pack and get my travel documents in order. I was flying from Ghana to Senegal and then to the United States on Air France.

I did not know there was a farewell party for me that evening. Around five o'clock, Nana told me to get ready because we were going to visit some friends of hers. I believed her until I arrived at a house that was packed with people. Mr. Roberts was there and the translators I had met. I figured it was a celebration of sorts but did not think I had anything to do with it. We had drinks and a variety of Ghanaian dishes. Everybody was having a great time, men seated on one side and women on the other. The men were loud, discussing soccer and politics. The women were notably quieter, either talking about family or some event that took place in the community.

After the dishes were cleared and it looked like we were winding down to leave, Mr. Roberts stood up and called everyone to attention. He told the crowd that he wanted to introduce someone in the community. He looked at me and asked me to stand. He went on to talk about my background and the possibility of my joining UBS. He thanked me for the visit and asked me to promise to return to Ghana whether I was hired by UBS or not. I had not seen this charming side of Mr. Roberts. I thought he was very guarded and that was part of his culture. They asked me if I had something to say. I tried to think of something to say, particularly about their hospitality, and then thanked Nana especially for taking me in as a sister and for all she did for me for the entire visit. I wanted to say something funny but hesitated because I did not know much about humor in Ghana and whether there were gender boundaries. I finished my short speech by promising to return as soon as time allowed. Then Nana stood up and said the women in the community wanted to give me a gift. She came forward and handed me a small box wrapped in beautiful gift paper. Everyone clapped and started singing in Akan, their local language. Nana turned to me and said, "They want you to open the gift now." I did and was awestruck by their generosity. It was a beautiful gold chain with earrings to match. Nana said they wanted to give me something that was authentic from Ghana and that her uncle, a goldsmith, had made it specifically for me. That made the gift even more special. It was the best evening I could have asked for at my destination and the end of my four-week trip on the continent. I told myself that I should never judge a book by its cover. My apprehension at the beginning and fear of how I was going to fare in Ghana was unfounded. It was colored by my visit to Cote d'Ivoire where the culture did not come as easily to me. I realized I had become comfortable with western culture and less so with the cultures that I did not know on the African continent. I realized I was an African by blood and soil but not by culture. There is no one African culture but African cultures. I needed time to learn about them to be able to embrace them.

Ghana was my last stop, and I was glad my tour had come to an end. I had learned a lot about myself, the continent, and its people through the prism of the few I met and interacted with. The adventure with UBS was an experience of a lifetime. I did not expect it, but it enriched my life through the places I visited, the people I met, and the knowledge imparted to me. I was cognizant of the fact that the opportunity was not in my cards, but I was nevertheless grateful for it. UBS spent a fortune on me, an experiment of theirs that allowed me to discover myself. This influenced my later conception of studying abroad in Africa, and I emphasized to the students that a trip to another part of the world is always a life-changing experience. All in all, I was ready to return to the United States.

— 14 —

Leaving California for Georgia

The UBS exploratory trip across Africa gave me clarity about the teaching career I had chosen. I realized that the classroom is where I belonged and if I had to do administrative work, the best would be to combine it with some classroom teaching. I decided I was going to continue teaching at Stanford University because the job with the United Bible Society was not a good fit for me. I wrote to the organization's director, thanked the organization for the opportunity to experience the excellent job it was doing, and expressed my regret that I would not be able to be a part of it because I loved being in the classroom more. I did not hear from them for about a month, but my colleague from Nairobi did. She regretted that I had turned down the UBS job, but she understood my decision. After a few more days, I received a letter from the headquarters accepting my decision and urging me to reach out to them at any time if I changed my mind.

With a clearer picture of my career path, I resumed teaching at Stanford in October 1987. I was starting my third year at Stanford University and getting comfortable with teaching, researching, and directing a national study abroad language program for college seniors and graduate students in African studies. I was the designated director of the Swahili program for that year with the responsibility of recruiting for and planning the trip, including the logistics for the summer program.

It was during this time, which was the last term before summer, when the idea of moving to Georgia was suggested. This was unexpected because I did not know anything about the University of Georgia. I had heard about the state of Georgia before I came to the United States because of Jimmy Carter who was well-known in Africa for his research on peanuts and treating river blindness.

This intriguing new development came one Thursday morning. I had a call from the main office telling me that two visitors were waiting to see me. I was not expecting any visitors that day but left my office to meet them. I tried to hide my surprise and proceeded to exchange greetings and introductions. They told me they were from the University of Georgia (UGA), and they were professors in the anthropology department. I will refer to them in this story as Dr. Ben Blount and Dr. Caroline Ehart. I invited them to my office where we talked. I learned that they were in San Francisco that weekend to attend the Annual Conference on African Linguistics (ACAL) and had come to Palo Alto to bring greetings from a mutual friend from linguistics circles. We spent about an hour talking about work and research interests when one of them changed the subject to UGA. They explained that linguistics was under the umbrella of the anthropology department. Dr. Ben was an anthropological linguist and Dr. Caroline was a biological anthropologist with extensive ecological research in the Udzungwa Mountains of Tanzania.

I was surprised by the next comments. Both professors said they were planning to stay in the Palo Alto area until later in the evening and wondered if I would join them for dinner. I gladly accepted their invitation but was surprised by it, since this was the first time we had met. After classes that afternoon, I went home early. I took time to locate the restaurant in downtown Palo Alto to make it easy to find that evening.

Dinner was at six-thirty, and I made sure I was there five minutes early. The restaurant was on the second floor of a hotel, so I took the elevator one floor up, and as soon as it opened, I saw Dr. Caroline at a table they had reserved. This was an incredibly nice and upscale restaurant. The professors had already ordered a bottle of wine, which was chilling in the bucket. For me, this ambience was special, more so because I did not eat out much, especially not in such a high-end restaurant, except for special occasions when I happen to be invited by Charles and Jean. Even then, this dining experience was exceptional.

The conversation was varied, ranging from academic to personal experiences in Africa. I learned that both had been in different parts of Africa for research. This made the evening even more interesting as we talked, laughed, and theorized on different linguistic and anthropological issues. After dessert, they revealed the real reason for their visit to Palo Alto. The University of Georgia had a strong interest in Africa. UGA had several faculty members across different departments with extensive research experience in Africa. The Department of Anthropology alone had five faculty members whose scholarship included this important research. The graduate studies program in their department had several students, both Africans and Americans, doing research for their thesis or dissertations on Africa. In addition to the faculty in anthropology and linguistics, the other faculty interests included agriculture, international development, geography, forestry and environmental studies, family and consumer sciences, and education. This interest led to the creation of an Africa interest group, which engaged in regular meetings in the hope of developing a campuswide African studies program/center. Dr. Ben explained that they were planning to apply for a Title VI grant to develop an African studies concentration, but the requirements included the study of African languages. They needed help in developing the language component and that was why they had come to see me. Dr. Ben was blunt in his proposal when he said, "Dr. M. thinks you are the best person to help us and that you might consider joining our faculty at the University of Georgia."

I was flattered and dumbfounded. One of the thoughts that came to my mind was the fact that I did not know much about the university nor the region. I had spent seven years in California and the only places I had traveled to, apart from Texas for a conference, were the Midwest and the Northeast. Dr. Ben was ready to persuade me to consider their offer. He declared, "I know you are going to be at the linguistics conference in DC next month, and UGA is willing to pay for you to travel to Athens after the conference for a visit and then connect your return to San Francisco from Atlanta."

Framing this opportunity as an informal invitation to visit made it difficult for me to refuse. There was nothing to lose by an offer to visit a place I knew absolutely nothing about. Furthermore, by agreeing to visit, I was not committing to anything. These fine professors left an important conference in San Francisco and spent an entire day in and around Palo Alto waiting to take me to dinner to ask me to visit their institution. It would have been unkind to turn them down cold. I clasped my hands like I was praying. I looked at both directly and agreed to their plan to visit after the linguistics conference in Washington, DC. I could see that they were relieved and happy. It was now almost 8 o'clock and I knew they were going to catch the train back to San Francisco and I had to teach the next morning. I thanked them for dinner, and they settled the bill. They then escorted me to my car and left for the train, which was a short distance from the restaurant.

Driving home, I started to wonder what this meant. A million questions were racing through my mind, many of them were "what if." I was not sure I was ready to leave California, my second home, for a completely unknown place with unfamiliar people, culture, and customs. I consoled myself by saying aloud, "This is only a proposal, and it will never come to fruition."

I decided not to mention the offer to anyone, at least not until after the visit to Athens. I also promised myself that I was going to try and block it from my mind and not give it too much thought until after the conference. I had a paper to write, and that was going to be my primary focus. Somehow, I was successful in doing myself that favor.

The meeting in DC went as planned. I had a presentation to make on the second day of the conference. Dr. Ben was there, and he knew I was traveling to Georgia the following day. I am not sure when he left the conference, but when I arrived in Atlanta, he was there with Dr. Caroline to welcome me. This was my first time at an airport where one also had to take a train within the airport to and from the gates. I thought Atlanta must be a special place.

The drive to Athens was interesting too. Compared to now, the landscape was more rural in 1988. Construction of highways and developments along them was negligible. There were farms with horses, cows, and sheep grazing. There was no Highway 316 at that time. The physical features were a welcome change from California, and I felt at home with the lush greenery and farms. As I approached Athens, I was greeted by big magnolia trees along the road and in neighborhood yards and azaleas around every building. The beautiful flowers gave me a special welcome.

We arrived at Ben and Caroline's house where I was going to spend the two nights of my three-day visit to Athens. When I arrived at their house, I realized something I should have noticed during our time together. The two professors were not just colleagues in the same department, they were a couple. Dr. Caroline helped me with my luggage and showed me my room. I settled in and freshened up, then joined them in the living room where Dr. Ben offered me a drink. Both started making dinner: spaghetti and meatballs and a green salad. This was complemented with fresh, sliced peaches for dessert, topped with vanilla ice cream.

After dinner, they walked me through my scheduled visit. I was surprised by the details. It started at nine in the morning and ended at five o'clock. Included were a visit with the vice president for Academic Affairs, the vice president for Legal Affairs, the dean and associate deans of Arts and Sciences, a visit to the main library, a luncheon with a select group of faculty and administrators at the Georgia Center President's Dining Room, a meeting with the head of the Anthropology Department and finally a meeting and reception with the department faculty. I was concerned about the meeting with the faculty because I had not prepared a presentation. Dr. Ben assured me that I did not need one because many people were familiar with my work and I could, if I liked, talk about my research for the paper I had recently presented at the Washington, DC, conference because it spoke to both linguistics and anthropology.

At the end of all these meetings, I had a good feeling about the University of Georgia. Everyone I met was warm and extremely friendly. Each meeting started with a description of the university, its strengths and, most of all, the long-term plans to diversify the campus.

I have a confession to make. I was not aware of the university's segregation history. I did not know Charlayne Hunter-Gault and Hamilton Holmes' story. In 1961, when they enrolled at UGA, I was in middle school. As children, we heard a lot about Dr. Martin Luther King Jr., but we associated his efforts in the United States with the struggle to rid South Africa of apartheid. During my visit, no one brought up the discussion about the university's history of segregation that would have made me connect the mention of "diversity" to the university's historical past.

Though I could not forget the institution's history, I had a good first impression of the school, the people I met with, and the general atmosphere. At the end of my meeting with the faculty and the reception, I thought I was just going back to my hosts' residence to prepare for my departure the next day. My flight was not until later in the afternoon, so I had time to relax or walk around downtown Athens. As if someone were reading my mind, one of the faculty members approached me and said, "I have been assigned to accompany you tomorrow morning with a realtor to look at houses in our community." "Oh," I said. "That will be nice." At that time, I had no idea what I was agreeing to or why I should be looking at houses. Shortly after this, the department secretary came to the reception area and said the department head would like to speak with me before I left. I followed her to the office and found the department head sitting at his desk. He stood up to welcome me again and commented on the discussion we had earlier about my research agenda. He continued to say that he would be happy if I would agree to join the faculty and participate in the plan to build the African Studies program at UGA. Before I could respond, he placed a letter in front of me and said, "You can take this with you, and if you think it is worthwhile, sign it and send it back to us." I took the letter and started reading it. When I saw the salary offer, I thought someone must have checked my salary at Stanford and wanted to be competitive to win me because the offer was three times the salary I had at Stanford. Additionally, I was offered moving costs, start-up costs, and a one-time research fund. This was too good an offer to refuse. I took the letter and placed it in the UGA folder given to me earlier in the day. I thanked him for a wonderful visit and promised him I would seriously consider his generous offer. I left the office and joined my hosts who were waiting for me to retire for the day.

Ben asked me how the meeting went. I am sure he knew what the agenda was, but he was anxious to know what I was thinking.

"Very well indeed," I replied. "This has been a very interesting and fruitful day, though exhausting."

I did not say anything about the letter, and they did not ask questions that would make me reveal my thoughts. Dr. Caroline suggested we find a place to eat before we went home. We went to The Last Resort, a popular restaurant in town. I wondered at the name and made a joke about it, "I hope this is not a place depressed and desperate people go to find a glimmer of hope." We all laughed and proceeded to get a table. The food was fantastic, and I was beginning to feel too comfortable in this place. After dinner, we headed back to their home, had some coffee, and then decided to retire. I had to pack that night because I was not going to come back after the Athens community tour. Before I went to bed, I re-read the letter. It was an incredibly good offer, and I was going to accept it. I took a pen and signed it, placed it in the envelope the department head gave me, and placed it on the dresser.

In the morning during breakfast, I announced that I would like to stop at the department head's office to say good-bye. I still did not mention the letter. In the office, I found the secretary who told me he was in class. I told her I wanted to say good-bye and return the envelope to him.

I left for my tour of Athens, which was interesting but overwhelming. The realtor showed me houses in the Five Points area on South Lumpkin and Milledge Avenue. These were areas within walking distance to the university and downtown Athens. She also showed me some property on the east side of town. In both cases, the houses looked huge, and since I had never owned a house, I did not know what to make of the experience.

After the tour, my host put me on the Athens airport shuttle to Atlanta, popularly known as "Door to Door," to catch my flight to California. Both the ride to the airport and the flight to San Francisco were smooth. Because of the time difference, I got home in the afternoon around four-thirty, West Coast time. As soon as I entered my apartment, it dawned on me that I did not have a lot of time left in this apartment and I had to start thinking about packing and moving across the country. Suddenly I realized the magnitude of my decision to accept the offer.

In Athens, word that I had accepted the position at the Department of Anthropology and Linguistics was spreading like wildfire. The news had traveled to Washington, DC, as well as to people from the Stanford-Berkeley Joint Center for African Studies who were attending an annual directors meeting. One of the UGA African Studies Program coordinators was attending this meeting and he received news from his department that I had accepted the position. In his excitement, he made a comment when talking to the participants about UGA's efforts to strengthen the curriculum by hiring me, along with the focus on Africa and the intention to apply for a Title VI Center federal grant. The director of the Stanford/Berkeley Joint Center for African Studies received the news with shock because he did not see it coming. Immediately, he called the Linguistics office at Stanford and told the office manager what he had heard. He also told her to tell me not to make any plans to go to UGA until he was back and had a chance to talk to me about it. The office manager took the liberty to let him know that there was a visit by two UGA professors a few weeks back and that she suspected that was the beginning of this new development. This made the director even more unsettled, and he decided to cut short his time at the meeting and come back to California to try to convince me to stay. When I showed up at the office in the morning, the office manager asked me about the rumor that I was moving to UGA. I asked her how she knew, and she revealed the frantic phone conversation with the director. She added that the director was upset about it and had requested that I make no decisions until he returned. I did not want to tell her that this was too late because I had already accepted the offer.

Around nine o'clock that same morning, the director came to my office to talk about what he had heard while in Washington, DC. He wanted to know why I wanted to leave Stanford for Georgia. He added, "The South is hard to live in. I do not think you are going to like it. There is the Ku Klux Klan, and racism is prevalent there."

I listened intently. I knew I had a choice and, if necessary, I could call Georgia and say I was sorry, and I had changed my mind. But I had not reached that point yet. The director continued to inform me that he had made an appointment with the provost to try and convince me not to leave. The appointment was at 11 a.m., and he added he would come for me a few minutes before then. For the two and a half years I had been at Stanford, I had not had an occasion to meet anyone in administration. I was not even sure who the provost was.

At 10:45 a.m., the director returned, and we walked across the quad to the provost's office. The secretary informed the provost about our arrival, and we met a young, very impressive Black woman, who introduced herself as Condoleezza Rice. Once seated, the director opened the discussion by thanking Dr. Rice for agreeing to meet with us on such short notice. He continued, "This is Dr. Moshi. She came to us from UCLA and is in the Department of Linguistics with major support from the Stanford/Berkeley Title VI Joint Center. She is a critical addition to our African Studies program, and we are happy to have her."

Dr. Rice turned to me and said, "We are happy to have you here. UCLA is an incredibly good school. So, I hear Georgia is trying to steal you from us. What is the offer?"

Before I revealed it, I thanked her for meeting me and said, "I would love to stay at Stanford. I have been happy, and I have grown a lot in my research goals in the two and a half years I have been here. Stanford has given me the opportunity to build my scholarship in both linguistics and African languages teaching and research."

I then proceeded to show her the offer letter from UGA.

"What? This is three times what you are making now!" she exclaimed. "How can they do that? It is above what we pay our associate professors here."

I wanted to give a response, which would have been, "That is Georgia." She turned to the director and said, "This is out of the question, and we cannot match. The most I can do is up the current salary by a couple of thousand and not more than that. Even then, we will have to split it with African Studies."

Again, I wanted to respond, but my cultural background made me bite my tongue. It is not culturally acceptable to go toe-to-toe with your boss. The director assured her he would discuss her offer with his Berkeley co-director and get back to her soon. He thanked her and she shook hands with both of us. I thanked her again for meeting us and we left. On our way back to the office, the director asked me what I thought of the proposal from the provost. I told him that there was nothing there to think about. He then vowed he was going to fight to retain me. I told him I was happy to hear that. However, deep in my heart, I knew this discussion was not going anywhere and I should focus on moving to Georgia. Later that afternoon, the secretary told me that the department head wanted to see me at three-thirty if I had the time. I agreed to the appointment, and I showed up at the scheduled hour. She was waiting for me. After a little chat about my work, she asked me about my intention to move to Georgia. I told her that was my plan. She paused for a moment and said, "From woman to woman, I would say to you, run before they cut your legs. No place makes anyone someone, but individuals build the reputation for a place."

This was a loaded statement, but I knew exactly what she meant. I needed to make the decision for myself and not because of something or someone else. She concluded, "I am happy for you, although we will lose you. All I can do is wish you all the best and be the scholar you are destined to be."

I was in tears. I thanked her for her kind words and advice. She then saw me out of her office and added, "Do not forget to stop by before you leave."

I thanked her again and left. I was glad that this meeting happened, and it was the last thing on my mind that day. The department head's words were assuring. My decision to leave Stanford, though tough because of the relationships I had already built, might be the right one at the time for both my career and life.

I had another month of teaching before the end of the academic year. I divided my time between teaching, grading exams, and packing. I also had to maintain communication with UGA in preparation for my move to Athens. It was overwhelming, but I thought it was manageable. I needed to complete packing before the end of June when I had to leave for Tanzania to direct a government-funded Group Projects Abroad, a monthlong program targeting graduate students and seniors in intensive language and culture programs.

I did not know much about companies that moved cargo across the country, which I could hire to move some of my belongings, including my car. We did not have cell phones and the internet then, so I could not do a Google search. We relied on the Yellow Pages, and I was lucky to find a company called Bekins. I later discovered that it had an office in Athens on Jefferson Road. The location was not familiar then, but because it said "Athens," I was encouraged. At the very least I could visit the office if anything happened or went awry. The other attraction was that they agreed to include my car in the same package and offered to come to my apartment to pack and move whatever I wanted to send to Georgia. With the car in the same consignment, I could pack as much as I wanted in it, lock it, and keep the keys with me. The price was good, $2,000, the amount allocated by UGA for moving costs. I made an appointment for the company to come in August, shortly after my return from the GPA program in Tanzania. I was expected at UGA in October, but I wanted to be there in September to find an apartment and settle in before school started.

I was glad that the Stanford Center for the Study of Language and Information (CSLI) wanted me to remain in their research pool because this allowed me to continue working with Dr. Joan who had developed a research interest in Bantu languages, specifically Chaga (my first language) and Chichewa. We still had some funds from the National Science Foundation (NSF), and I had a great deal of data to transcribe from the collection the year before. Remaining on the project gave me an opportunity to travel to Stanford each summer for the next three years after I left to complete the NSF-funded project.

*Visiting in 2018 with Winnie Lee (left) and
Vera McIntosh (middle), my 1978-80 host family
during holidays when studying in England*

*Jean and Charles Prael, Palo Alto.
They supported me while in graduate school at UCLA.*

Charles and Jean Prael (with me in the middle) in Palo Alto, 1984.

— 15 —

Up the Ladder at the University of Georgia

Many of my friends and students at Stanford University were not happy when they found out I was leaving California. My American adoptive parents had mixed feelings about the South, but I assured them that I was going to be fine. My life was an adventure, I told them, and this would be another chapter in my journey. Besides, I could always come back to California for breaks. Between May and the end of June 1988, I had endless farewell parties. Finally, my seven-year stay in California ended the first week of September when I left for Georgia.

My first week in Athens was smooth. I already had a good relationship with Ben and Caroline from the first visit. They offered to host me again until I found a place to live, which took a week of apartment hunting. It is always tricky when moving to a new place to decide on a suitable location. I wanted to find a home that was close to shops but not too far from the university. Caroline lived in Athens before she married Ben and relocated to Watkinsville, Oconee County. She advised me to work with a rental agency and suggested Delta Realty, whose office was at the junction of Gaines School and Barnett Shoals roads. The staff at the Delta office was extremely hospitable. I discovered later that they had experience renting to international visitors. The properties they managed were all over Athens, but the manager suggested, based on my preferences, their duplexes on Gaines School Road. We left the office to look at two units that were available. Both apartments were empty, allowing me to make a choice. When driving in, the duplex faced the main road, so I chose the apartment on the left because it had a bigger parking spot and a space to create a garden for flowers or vegetables. The other nearby units had tenants who gave me a sense of being in a community and not isolated. The neighbor on the right had two little children, a boy, and a girl, approximately age seven and four. The only thing that worried me was the fact that the back of the house had undeveloped land with very tall trees. During the spring and summer, it was hard to see the houses beyond the tall trees.

Despite my fear of the bushy landscape in my backyard, I liked the unit. It had two bedrooms, a living room, a kitchen with a breakfast area, and storage space. This was more room than I ever had in California. The rent was also a deal, $240 a month. I was blown away considering that when I was at UCLA, I paid $500 for a one-bedroom apartment, and in Palo Alto, I paid $750 for a similar one and $1,000 when I moved to a two-bedroom apartment. I agreed to rent the unit and we went back to the office to sign the lease. I emailed many of my friends in California about the low rent, advising them to relocate to Georgia and save their money for retirement. This was true because after two years of saving, I was able to buy my first house for a price one could never dream of in California.

After signing the lease, I started setting up my new residence. Instead of immediately buying all the furniture I wanted, I decided to rent until I was ready to buy. I needed time to discover Athens and find places to buy good furniture. This turned out to be a good decision because I got the furniture I really wanted when I bought my first home.

Coming to Athens early was beneficial. By the time school started in October, I had everything I needed in my new place. I had familiarized myself with the neighborhoods and knew where to find the essentials. I also had my office set up and attended orientation meetings offered by different units to new faculty. I met a lot of faculty members of color, which convinced me that when the 20th president of UGA, Charles Knapp, promised to bring diversity to the university, he meant what he said. President Knapp is considered by many folks to be the most successful in this mission during his tenure (1987-97).

One of the agreements I had secured in my contract was to use the fall term (October-December) for research and teach in the winter (January-March) and spring (April-June) terms. With that time, I was able to settle in and get ready for my classes in a new environment. I had an opportunity to sit in on some of the classes to get the feel for the interactions between faculty and students. I also had a couple of invitations to visit classes where I spoke on a topic of the instructor's choice. This was valuable experience prior to my first term teaching at UGA. It was also a good time to make myself available for African Studies meetings and to get to know colleagues who taught in different units on campus and to hear about their interests and experiences in Africa.

My first class in winter term was like a blind date. I was very prepared for the class but even my extensive teaching experience and a distinguished teaching award from UCLA did not ease my nerves. The male students dressed in business casual, which reminded me of the Stanford students, but was unlike the very casual UCLA students. None of the male students wore jeans, designer or otherwise. The female students were a different story. Most of them were dressed like they were going to a formal party, every hair in place, makeup, and high heels. It was cold, but many of these dresses were sleeveless. They had a jacket or cardigan on the chair or a winter coat for when they left class and went outside. Thank God the heating in the classrooms was good, sometimes too hot for me in a cardigan or jacket. Based on this group, I was underdressed, but I was trying to look professional. I was their professor.

I started the class as I would at UCLA or Stanford. After a general greeting to all the students, I asked them to stand up and walk around the class. They were to say hello to at least five other students and try to remember as much of what they said to one another as possible. I, too, left my desk and podium, and walked around the class saying hello to some of the students. My intent was to break the ice and have the students learn from the get-go that they were in this class to learn together and that my method of teaching was less teacher-oriented and more student-oriented.

After the five minutes I had allocated for this portion of class, I brought the students back to their seats. Randomly, I picked students and asked them about who they had met, their names, where they came from, what they were interested in, and so on. Very few students, other than the ones who were acquainted already, had much to say about the person who they had met for the first time. But my goal was met, and the ice was broken.

When this exercise was completed, I told them a little about myself, that I was born in Tanzania on the slopes of Mount Kilimanjaro, I went to school in Tanzania where I obtained two degrees, a BA in linguistics with a minor in education and master's in linguistics. I then went to England where I was hoping to get a Ph.D. but was forced to do another MA before I could advance. Due to financial hardships, I had to leave, and that is how I ended up at UCLA where I was determined to get a Ph.D. in the shortest time possible. I was able to complete my final degree in four years and two months and was later employed by Stanford University. After three years, I was recruited by UGA, and the rest is history. I could see that some students were listening curiously to what I was saying. As soon as I finished, a few hands went up. One student asked, "What it was like to grow up on the slopes of Mount Kilimanjaro?" The student then declared an interest in climbing the mountain someday because he heard it was the highest in Africa.

Although it was important to make the first class simple by developing a comfortable learning environment, this also was an important question in the cross-cultural exercise I had planned for later. So, I was not sure this was the right time to delve into my life in Tanzania. I tried to be diplomatic and reassure the student that what they asked was very interesting, but I was going to devote a day to cultural discussions when I would share my background and answer all their questions about me and Kilimanjaro. I could see he was satisfied with the answer, and I was relieved. The second question came from a female student who had a mirror in one hand and lipstick in the other. She started by saying, "I do not like linguistics, and I am taking it because it is a requirement. How can you make me like it? It is not important for me to be in school. My parents made me so I can find my future husband here."

My heart sank to my feet. I knew I had a big problem motivating students like her to do well in class. The course was Introduction to Linguistics, which we sometimes call Baby Linguistics, so I quickly formulated an answer to try and calm her nerves as well as others who might be feeling the same.

"Okay," I said. "I will try to make it fun. I am glad you have a mirror in hand because, at some point, I am going to require everyone to bring one to class to see what our mouths and tongues do when we speak. We are also going to learn why we do not all speak the same way and how the languages of the world are related. I am sure you would like to know that and share that with your future husband, and your children and grandchildren."

From her facial expression, I could see she liked my answer and quickly put her mirror back in her purse. Spoiler alert: This student became a linguistics major in the end and continued to a master's degree. She took a couple more classes with me and because she liked psycholinguistics so much, she joined law enforcement to do forensics. I wonder what she might be doing now and whether she did marry a UGA graduate to please her parents.

The class was only fifty minutes, and I had a few more housekeeping things to do before the students left. I distributed the syllabus and talked at great length about the requirements and responsibilities—especially attendance, homework, no make-up policy, the culture of honesty, and the importance of using the office hours. By the end of this, time was up, and I could see the students putting their gear back into their bags, ready to take off. Unlike in Tanzania and England, most college students in America do not wait for the teacher to dismiss the class. I learned quickly to budget my time to make sure I finished the lesson right when class was scheduled to end. But sometimes the students would start packing up five minutes before the end of class and it did not matter that I was still talking.

I was ready for my next class, which was Swahili. I was not sure what kind of students would show up since it was the first time it was offered. The class had been very well advertised around campus and on department bulletin boards. I had a five-minute break to go to my office in the basement of Baldwin Hall, home of the Anthropology Department. I had just enough time to sip my tea from a thermos, grab my Swahili material, and head upstairs to class.

Interestingly, I had only ten students. All but two were graduate students in anthropology or linguistics. The two undergraduates were from the Religion Department. I followed the same orientation as I had in the linguistics class. Initial greetings followed by students greeting one another and then introducing myself. During this time, I learned more about the undergraduate students. Their parents had been missionaries in western Tanzania, in the small town of Kigoma, which sits on the border of Tanzania and the Democratic Republic of Congo. They were both born in Tanzania and had learned a little bit of Swahili from their nannies and playmates. The two students were interested in continuing to learn the language because they wanted to go back and continue the work their parents had started. The two did not know each other before class, but they quickly forged a bond that helped motivate their learning. Though I would love to know where they are now, I lost touch with these students after they graduated from UGA.

After distributing the syllabus, my plan was to teach the students a couple of songs to help them learn the alphabet and to count. Using songs as a tool for learning a new language has always been one of my techniques because it keeps students alert and excited to continue learning. This method also became a recruitment tool, highlighting such a rare language as Swahili on a campus like UGA to large numbers of students. Word of mouth from the students in that class was the catalyst for the soaring numbers of students that joined my winter term beginner class.

Incorporated into my teaching of Swahili language and culture was a cultural awareness program that I had started at UCLA. I designated one day in the term when students would focus on Africa's myriad cultures. Just as I had at UCLA, and now at UGA, I took charge of the first event by cooking a variety of African dishes and invited all the students to my house. Their contribution was a non-alcoholic drink of their choice. As a group, we discussed the different types of food, cultural aspects associated with it, and the general division of labor in an African household—who provided the food, who prepared it, who cleaned the dishes, the role of older children in the household, and the age of involvement. The students were surprised when I told them that I made my first family meal at age five. The pot was too big for me to remove from the three stones that held over the fire logs. I had to be creative. My mother was still at the farm, and I needed to feed my siblings. So, I removed all the hot embers and served the food from the pot still sitting on the fire stones. It turned out I had just created a food-warming solution. Because it was still hot around the pot, the food remained warm for hours until my mother and the rest of the family came home later.

Future cultural awareness programs were designed differently where groups of students were given recipes and asked to prepare a specific dish. Each group described their dish and how it was made. Some of the dishes were a success while others were a miserable failure with preparers warning their fellow students ahead of time that their dish might not be edible. Through the trials and tribulations of making the assigned dish, the students made the most hilarious and merry memories.

My favorite story came from a group that was supposed to make curried rice with coconut milk. Instead of buying canned coconut juice, the students decided to strive for authenticity and do what they had seen in the Swahili video in class, which was to buy a real coconut, crack it, grate it, and make the juice. The students lived in one of the high-rises on campus. On the day of the event, they started early to make their assigned dish. But they ran into a problem when they could not crack the coconut. They tried different ways, but it would not crack. One of the students in the group suggested dropping it from the eighth floor of their hall to break it. When one of them went to retrieve it, he found nothing had happened to it. Frantic, since time was ticking, the group decided to call and give me the bad news about their assignment. Instead of rice with curry sauce, they had decided to do rice and cabbage because they could not make the coconut curry. When I asked what had happened to the coconut, they sheepishly said they could not crack its shell. I asked where the coconut was, and they told me it was on the kitchen counter. I told one of them to hold it sideways on the palm of the left hand. Then I directed them on how to crack it with the blunt side of a knife. "Hit the shell lightly while moving it around the palm of the hand and watch the coconut as it slowly starts to crack," I said.

I could hear them marveling at how simple it was. However, this was not the end of their problems making this dish. Though they cracked it, they now did not know how to extract the coconut meat from the shell to make the milk.

"Okay," I said, "use a knife and scoop it out, put the milk in a blender, add hot water, and then put the blender on grate. Then, put the grated coconut and the juice through a sieve or cheesecloth and squeeze. Add more warm water until you get enough juice for your sauce."

After they listened intently to my instructions, the drama ended, and they had a perfect curry.

Another memorable incident was when students were assigned to make an authentic African fruit salad. After searching the web, they went to different grocery stores looking for tropical fruits. They found papayas, mangoes, bananas, and pineapples. Cutting up the fruit was going well until they came to the mango. They tried to cut it and hit the hard seed inside. This was their first time cutting a mango and they could not figure out how to proceed. They decided to bring the fruit salad to my house, where the event was being held. They were smart to come a little early to finish preparing it. They were surprised when I showed them how to cut the mango around the hard seed. Their eyes lit up, and they started laughing. None of them had any idea that a mango had such a big seed inside. They confessed they had never bought a whole mango, only fruit cups with sliced mangoes.

During these cultural awareness classes, I anticipated mishaps like these. To be on the safe side, I had a backup plan—make extra food and desserts. Also, the backup plan came in handy when a participant showed up without the assigned dish for one reason or another or someone brought something no one could eat. Nevertheless, the program was valuable for helping students learn about another culture from experience. Many of them repeated the assignment at their convenience and shared their success or failure with the class. Whenever I had time, I invited them to my house to watch me make the dish instead of reading a cookbook. Many perfected their skills over time. Once, a student brought a dish to class to prove that though her first attempt failed, she ultimately succeeded.

The final project was a group assignment, a report on a cultural event that emphasized the lessons learned and skills acquired. The best lesson was cultural understanding, the most important ingredient in the recipe for world peace and understanding.

The students' response to the cultural assignments gave me hope and kept me excited to be in Georgia. The students' enthusiasm has allowed the African language classes at UGA to thrive for as long as they have when other schools continually struggle to keep their programs alive and off the chopping block from budget cuts. I was thrilled when, in my second year of teaching the language at UGA, the enrollments rose from eight graduate students and two undergrads to twenty-five students, a mixture of graduates and undergraduates. The program was safe and would have a chance to prosper.

Teaching enthusiastic students was a big motivation for me my first year. I struggled to understand Southern culture. Despite what my colleagues in California had told me about the South, I was confident that the culture would be closer to African because of that country's influence here. African slaves had been brought to the South, and because when I had visited, I was shown kindness—popularly known as Southern hospitality— that resembled the African way of relating to others. I also had found foods that were typically African, such as collard greens, plantains, green bananas, yucca, and taro root. I realized that these also were associated with Pacific Islanders, but they too were highly influenced by the African coastal cultures. Looking back, I think I may have been a bit naive because cultural interpretations are not exported wholly and can be influenced by individual preferences. I had assumed that every Black person would interpret cultural experiences the same as me, but I was proved wrong by my first encounter. An African American administrator invited me to her home for afternoon tea. I was excited but did not know that two other African scholars who had been hired before me had also been invited. When I saw them, I thought it was a "welcome to UGA" gesture for me since the others had already been at the university for a year. I was also interested in meeting the administrator because I had heard that her area of expertise was education in Africa and that for her doctorate dissertation research, she had gone to Tanzania.

The tea was scheduled for three o'clock and I arrived in a timely manner. Shortly after I arrived, I was joined by the two African scholars, one from the Democratic Republic of the Congo and the other from Sierra Leone. We were warmly welcomed and served tea and cookies. We had time to discuss cultural and academic topics. I answered a few questions about Tanzania and our host told us about her research experiences there. She had both positive and negative experiences, and I attributed the negative ones to a clash of cultures. Often, we make assumptions about the treatment we will receive from people who look like us and then we are disappointed when our assumptions are not met. I have seen this happen many times with my African American students in my study abroad programs. They often assumed that when they were in an African country, they would be received with a "welcome home" attitude. They were disappointed when they were treated like any other visitor. I understood their frustrations and disappointment because I have been in their shoes. When I went to West Africa for the first time on my United Bible Society exploration, I assumed I would easily fit in, and the people would accept and embrace me as one of their own because I was African. I quickly realized that not only was there tribalism, but there also was nationalism. I was as foreign as any person from Europe, the Americas, or Asia. Color was never the determining factor for who was the "other." Color was used in a different cultural sense, like a measure of beauty in women. Light-skinned women were deemed beautiful while dark-skinned women were not. During my second visit in 2012, unfortunately, I did not find much improvement in the attitude, but at least I knew what cultural mistakes I should avoid. Learning from these experiences informed me on how to prepare students embarking on different programs in Africa. Intensive orientation sessions were a necessity to moderate my students' expectations and minimize their disappointments.

With these revelations in mind, the memory of the afternoon tea party was significant for my understanding of different expectations. I was not prepared for what our host would say after describing some of her unpleasant experiences in Tanzania. She asked me why I chose to come to the University of Georgia and decided to leave an Ivy League school like Stanford. I told her that I liked what I saw when I came for the visit in the spring, and I was confident that I had more room to grow professionally at UGA. To this, she responded, "Yeah, that is right. You people come to schools like UGA knowing very well that the administration will pit you against us and get away with it. You are Black, we are Black, who can tell who is who? There is no distinction between Africans and African Americans."

I was dumbfounded. I looked at my fellow guests and they had their eyes fixed on the ground. I am glad I did not have my cup of tea in my hand. It was fine china, the likes of a European tea set. If I did, I am sure I would have dropped and broken it, and that would have made matters even worse. The only thought that came to mind was to find a way to escape this tense situation. I waited for someone to say something to break the silence, but everyone was quiet. I summoned up some courage and said, "Thank you for inviting us to your house, it is lovely. Thank you for the tea and cookies too. Very lovely. I do not want to break up the party, but I must leave now because I am going to the evening Mass at the Catholic Center, which starts soon. Thank you again. Hope we can do this again at another time."

I picked up my purse and she walked me to the door. I got into my car and drove off to the Catholic Center. I had about fifteen minutes before the service started so I had time to sit in the car and evaluate what had happened. Although I did not understand it, I told myself that this was just one person's opinion. I was concerned because she was a member of the African Studies interest group, and I knew I would see her many more times in meetings and other gatherings. I decided that I would have to take each day as it came. I thought about my experiences at Stanford and realized that I had not experienced the diaspora culture. Indeed, at both UCLA and Stanford, I was isolated from this culture. I had no experience at all, and I did not know enough about the African diaspora, its history, philosophy of life, belief systems, and experiences that included oppression and racism. I had assumed that my experience was like that of African Americans and that I would be embraced through and through. I told myself that my work was cut out for me, and I could not expect anyone to change to accommodate me, but I had to change internally to accommodate others. As such, this incident during my first term at UGA left an indelible mark that helped me develop a platform to operate from, to be accepted for what I did and not for who I was. Focusing so much on who I was would not help me move forward nor successfully shape my future. I was glad to find that the incident was an isolated one. The road was going to be rough because I had to navigate three different cultures and outlooks on life: Tanzanian (African by soil and blood), African American (African by blood), and American (basically white). I told myself that nothing was impossible, and that optimism is better than pessimism.

My second encounter with racial intolerance was when I was trying to buy my first house. To be in the United States and to have a home I could call my own was both a joyful and painful thought. Where I come from, one builds a house. Until recently, no one bought a ready-made house. Building a home can take between five and ten years to finish. One saves some money, buys the material, and builds until the material is gone. The cycle is repeated until the building is completed. Most people move into the unfinished house once they have windows and doors installed and continue building the house while living in it. In the end, the owner does not owe anybody because it is a "pay as you go" situation. In America, the process was very different but also seemed simple: Get a loan from a bank, sign a contract for twenty-five or thirty years, and move into a finished house. The only difference is that you owe that money and must pay it per the contract. What most people do not realize is that, by the time one pays off the debt, he/she would have paid more than the actual sale price of the house. But there is no alternative unless you are wealthy and can pay cash.

Although the loan idea was unsettling to me, the thought that I would own my home convinced me to go through with the process and worry about the results later. I talked to a friend, Barbara Ferré, who was, at the time, a graduate student in UGA's Linguistics Department and a student in my Swahili class. She needed the language for her graduate research requirements. Barbara was one of my first graduate students in the Swahili classes and she did so well that she became my first teaching assistant with assignments in my first-year classes. I think her success can be attributed to her experience in learning foreign languages. She was born in Germany and spoke English, French, and other European languages. Her husband was an American and a professor of philosophy. I was a regular dinner guest at their house, and I benefited from talking to her husband about his knowledge and experiences at UGA. He taught me how to build my scholarly portfolio and how to position myself for success at this institution. I knew that Barbara would have some good advice from her experience buying the house in which they lived.

Barbara gave me the name of their Realtor, RE/MAX, and made the appointment for me. Two days later, I received a call for my first appointment with the Realtor. I did not know what to expect but was open-minded about learning from this new experience. After a few minutes at the RE/MAX office with the agent who asked questions about my ancestry and income but nothing about the kind of house I was thinking about, she invited me to ride with her to look at houses. We drove around Five Points, an area near campus with expensive houses, to look at those listed for sale. However, we did not get out of the car to look at any. She reminded me that I would not be able to afford any of the houses with "my" kind of money. I was not sure what she was talking about, so I did not respond. I decided to just listen. After about half an hour we left Five Points and headed to the west side. I was not familiar with Athens' neighborhoods, but now, upon reflection and after years of living in this town, I know where we were. We crossed the railway tracks, heading toward a downtown Athens neighborhood known for housing projects. I saw a lot of homes close together, some with unkempt lawns, and some that needed major repairs. She turned to me and said, "I think you might be able to afford some of the houses here." Again, I chose not to respond.

"Would you like to see any of the listed houses here?" she asked.

I had no choice but to answer. "I think I would like to drive through the town on my own first and once I decide on the location, I will take the next step."

I think the Realtor had allotted an hour for the first day. Then she added, "We will go back to the office so you can pick up your car, and perhaps tomorrow I can show you some listings on the east side of Athens."

I concurred and thanked her for her services. Before I drove off, she asked me if there was a convenient place on the east side to meet. I chose the Kroger parking lot, which was close to where I lived at the time. We then parted.

The next day of house hunting went the same as the first. The Realtor advised that it might be better for me to wait at least five years before I dreamed of buying a house because "my" kind of money would not do it. Her statement continued to puzzle me because I was not sure what money she was referring to since I only told her what my annual salary was but did not tell her that I had other savings. On the east side, we looked at houses near the Athens Airport. There were some nice houses, but they were too small. Others were unkempt and the yards were overgrown. I started growing impatient and uncomfortable with the whole exercise. I decided to make my voice heard.

"I like your advice about waiting on buying a house. You are right; I will wait for now. I do not want to waste any more of your precious time as I am sure you have a ton of customers waiting for your help."

She kept quiet for a few seconds and then said, "It's okay. We can look for as long as you want."

"It is enough for now. We will try again another time, thank you," I said.

We headed back to the Kroger parking lot where I had parked my car. Before she left, she shook my hand and said, "When you are ready, call me. I will be more than happy to help."

"Thank you very much," I responded.

Before I headed home, I went into Kroger to do some grocery shopping. As I was leaving the store, I saw a pile of booklets with house listings, both to buy and to rent. I picked one up. At home, I flipped through the pages and to my surprise found a listing on the east side that looked good. The listing was under Fidelity Real Estate. I called the number and a woman picked up the phone. I was surprised when she asked me where I was because she wanted to come out and show me the house. It took about thirty minutes, and she was at my residence. She introduced herself as Linda and offered to drive me to the house I had identified in the Green Acres subdivision. When we got there, we were able to go inside and look around. It was nice and only about six years old. I was most impressed by the Jacuzzi-style bathtub, but the house had two bedrooms, and two bathrooms. I did not think this house was big enough, particularly with the lack of storage space. Then Linda informed me that there was another house that had just gone on the market and was not in the booklet listing. She could show it to me if I liked. I agreed, and we drove to it.

The house was in the University Heights subdivision of Clarke County. We drove to the house in under three minutes from where we were. Compared to the other areas I had seen the previous day I liked the neighborhood. The house was about two miles from UGA and within walking distance of many shopping areas. I felt this might work. This house was close to College Station Road, the main road that runs from East Campus Road and connects to Barnett Shoals. Once onto the main drive, we turned left into a driveway where the house stood. There was no one home and the agent opened the large, black padlock on the door. I thought it was strange that we were entering someone else's house when they were not there. But this was America, and to me, everything was strange. We looked around the house—bedrooms, dining room, kitchen, and an unfinished basement. We came back upstairs, and the agent proceeded to open a side door leading to an open deck with a barbeque that was partially covered. From the deck, we could see the backyard, which was big enough for another house.

"What do you think?" she asked. Although I initially thought it was too big, I told her that I liked this house. She disclosed the price the seller was asking: $75,000. She then qualified her statement by adding that the owner wanted to sell quickly and was willing to go down on the price if the buyer was ready to make an offer. The seller was a widow with grown children who lived in Nashville, Tennessee, and she wanted to move to a smaller house. In fact, she had already bought her new home, which was her incentive to sell quickly and leave this larger house behind. Linda thought we could offer $72,000 and have the current owner pay the closing costs.

Although I was afraid this decision was moving too fast, I allowed myself to be captured by Linda, mostly because she was persuasive and treated me as an equal. From the time we met, we connected, and before we parted, we seemed like we had known each other for a long time. Despite the great location and other amenities that were beckoning me to take a leap of faith, I was still afraid to commit to the house.

Linda asked, "So, what is on your mind? Do you want to look at some more or do you want to consider this house?"

I thought for a second and then responded, "I will consider it if you can get the price down."

"I will do my best and confirm with you by the end of the day," she said.

Linda kept her word and by three that afternoon, I received her phone call confirming a price drop. She then asked me to meet with her the next day, Sunday, at 3 o'clock, to sign the preliminary agreement for Fidelity Real Estate Company to represent me in the buying process. She also informed me that I had to bring a check for $1,000, which was "earnest deposit money." This was another term I was not familiar with—money one pays to a real estate company to retain them. I learned later that the $1,000 was considered standard by all agencies and once paid it was nonrefundable even if the deal fell through, which made me extremely nervous. I needed to think seriously about what I was about to do before I issued the check. I felt like I was on a roller-coaster. The events of the day were overwhelming, to say the least.

Later that evening, I called a friend of mine, Nthwana Mzamane. She was a friend of mine who was South African and married to a colleague in the Comparative Literature Department. I wanted to show her the house and get her opinion. I expected her to tell me I was crazy so I could exit the process without the guilt or fear of losing my deposit. She promised to come around ten the next day. Exactly at ten, I heard a knock on the door, and it was Nthwana. We did not waste time but drove to the house, taking a few minutes to cruise around the subdivision first. She wanted to get the feel of the neighborhood before she could give me an honest opinion. Finally, we arrived at the house and parked on the street. With the black padlock on the door, we figured there was nobody at home. We walked down the driveway so she could see the backyard and get a pleasant view of the outside of the house. She was impressed that it was brick and not frame. I told her that was one of the features that attracted me to it. We walked back to the car, and I asked her what she thought. She said, "Go for it."

"But it is too big, I should look for a much smaller house," I declared.

Nthwana looked at me and said, "You are a real fool. There is no house too big to fill, and it is costly to expand a small house. We are Africans; the house will always be full. This is not too big."

I did not expect this response from my friend, but I was glad to get it, for she turned out to be right. Her words assured me that I was not foolish in purchasing a new home although this was the craziest thing to do in a foreign country and in my third year in Georgia. One of my colleagues, a fellow African who was in the Anthropology Department, had told me I was making a mistake to buy a house before I got tenure. Luckily, he told me this *after* I purchased the home, otherwise, his advice could have influenced my decision. But I resolved that if in three years I had to leave Georgia, I would rent it until I was able to sell it. I had gambled and now I was in it to stay the course, with the support of my friend Nthwana.

Later that Sunday, I went to Linda's office as agreed. The Fidelity Realtors' office was on Lumpkin Street, across from East Rutherford. At the office that afternoon, I had two things to do: sign the commitment contract and pay the earnest money deposit. Once settled, Linda advised me to visit the bank as soon as possible to initiate the loan application process. I went the next day, which was Monday. It seemed like everything was going well and lining up for my newest venture.

But I had surprises that were not necessarily comforting. First, I had an unexpected exchange with the loan office after I had already completed the better part of the loan application form. The agent wanted to know when "we" would be coming back to sign the application. I was not sure if he was using the inclusive "we" or if he meant someone else from the realty company had to be here that I was not aware of.

"Who are "we?" I asked.

"Oh, I meant you and your husband," he answered.

When I told him that I was the only one signing the application, he seemed disturbed. He asked if there was a problem with my husband being a co-signer. I asked him why it was necessary, and he indicated that it was not, but it would be unusual for the wife to make the application without the husband. I wondered whether it would be acceptable the other way around, but he had already told me that my case seemed unusual. I toyed with the idea of telling him that I was single, but I did not know how he would react to that, and I was afraid I might jeopardize my chances of getting the loan approved. I decided to assure him that it was not a problem in our family for the woman to be in charge and that this time, I would be in charge. That boldness took him by surprise. He went back to the application form and selected the "single" box in the marital status section. He told me the waiting period was five to ten business days and if I did not hear from them, I should call or stop at the bank to inquire after my application. He gave me his business card and escorted me to the door. He shook my hand and thanked me for coming in. I left the loan office conflicted and worried I was going to lose my deposit because I might not get the loan.

When I got back to my office, I called my friend Nthwana and told her what had happened. She was just as surprised but assured me it was unlikely the bank would deny the loan because of my marital status. Nthwana was correct. Three days later, Linda called to let me know that the bank had approved my loan and I should hear from them soon. She wanted me to visit with her as soon as I heard to walk me through the remaining portion of the process. Sure enough, the bank called the next morning. I spoke with Linda that afternoon where I learned what to expect at the closing date, the scheduled building inspection, and many other small steps before ownership of the house was officially transferred. It took about three more weeks to complete the process. The closing date was in the middle of May. This was timely because I had a trip scheduled to attend a workshop in Geneva in June. I wanted to move into my new house before the workshop because after my return, I was going to fly to Stanford for my summer research contract at the Stanford Center for the Study of Language and Information.

During the last part of the closing process, I received the biggest shock while at the bank. I learned how much the bank was charging for the different fees including the "points," a real estate term that refers to the fees paid directly to the lender at closing in exchange for a reduced interest rate. I realized I needed to write a check for $9,060, which was all the money I had in my savings account. At the Fidelity Realtors' office, Linda had done some calculations for me, and I was under the impression it was going to be a little under $6,000. Part of the increase was the fact that the bank had set my interest rate at 10 percent. I could not help but think about my experience filling out the loan application. I did not want the banker to bring up the spouse issue again, so I was determined not to let them see me sweat. After signing all the paperwork, I wrote the check and picked up the folder with copies of the documents ready to go. I shook hands with the owner, the Realtor, and the other bank officials. Congratulations were flying, but I barely heard them. I was focusing on what I had just done. I drove home despondent and worried about how I was going to manage the three summer months with no income. I knew the first mortgage payment was not until July, but it was still a long way to the first paycheck during the first week of October, since UGA was still on the quarter system. School was back in session after the summer in September. I knew I had to do serious budgeting to make it work.

For a week, I did not want to talk about the house to anyone, but I called my adoptive parents in California to share the news. When my friend, Nthwana, called to ask about the closing, I told her that all went well, and I was waiting to finish exams to start packing to move. She wanted to see the inside of the house, but I was not ready to deal with it until I was done with exams. My adoptive parents, on the other hand, were happy that I had made such a big step so early in my career and praised me for it. I did not discuss what had happened at the bank because I knew they would be upset about it. I was surprised when I received a congratulations card in the mail with a check for five hundred dollars as a housewarming gift. I was again impressed by their generosity especially now that I was fully employed at a prestigious university. The gift was more than welcome because it gave me some hope for surviving the summer, which was also supplemented by the stipend I was going to receive from the Stanford summer research program. I was also surprised when I arrived in Geneva for the workshop to find that my participation was highly valued and that I had a daily stipend for the entire seven days. The organizers also reimbursed my ticket and paid my hotel expenses. Before I left for Geneva, I took out a $2,000 loan with my car as collateral. This gave me enough of a security blanket until my first paycheck in the new academic year. As soon as I returned from California, I paid off the car loan to reduce the unnecessary interest that was accruing.

Buying a house in the United States at that point in my life was a teaching moment and a great experience that I willingly share with others. To this day, if I were to make a major purchase, I still consult and bring in an outside source because an eyewitness may affect other peoples' attitudes. And what's more, rather than relying solely on what others say, I verify the information before making any decisions by reading legal documents and instruction manuals. I do this no matter how long it may take or how confusing the legal rhetoric may be. I also remind myself to read between the lines and ask questions.

Although my belief system was tested through these early personal experiences and many more times professionally in academia and administrative work in the years that followed, it did not affect my view that everything we encounter in life is not meant to destroy but rather empower. I learned how to manage each event by focusing on acting rather than reacting. At UGA, there were times I doubted myself and questioned my decision to move from California to Georgia. This self-doubt was clear during my first year following an incident that happened in one of my classes.

Halfway through one of my linguistics classes, I noticed there was a student at the back of the class who had his legs on the desk and was reading the student paper while I was lecturing. I kept talking while I walked toward his seat, and he continued to read. I stopped and asked him to put his feet on the floor and to put the paper away. He looked at me and then loudly laughed. All the students turned to look at what was happening. As I was walking back to the front of the class, he shouted, "A professor is a 'he,' not a 'she.' Besides, you need to go back to Africa and teach your kind."

I ignored him and continued to teach. Luckily, I had only a few more minutes until the end of class. I gave the students a reading assignment for the next day before dismissing them. I stayed in the room until everyone was gone except the student who had been rude to me. As he was leaving the class, I said to him, "Do not bother coming back to this class. You can find a 'he' professor for yourself."

He gave me an angry look and exited the classroom without a response. I was glad and thought later that it was a good thing I didn't say anything else to him because the situation could have escalated. One thought that came to mind was that although I was not a male professor, I was the only female professor who was not born in America, but rather, in Africa, on the faculty at UGA. I was the first one too. For me, my status was quite an accomplishment.

Since this was my last class for the day, I decided to talk to the department head about my experience in the classroom. Fortunately, he was in the office and was willing to see me without an appointment. I told him what had happened as well as my decision. He looked at me and said, as a matter of fact, "You cannot throw a student out of the class. You just must bear with him and help him learn." I wanted to shout, "Whose side are you on?" but I did not, so I simply thanked him for his time and for listening. I left the room and went back to my office. I thought about my experience as a student in the British system. An incident like this would have been treated very seriously. I realized I did not have enough power to control what goes on in my classroom. I started to doubt myself and question whether I could survive in this environment. I had not developed a network of trusted friends yet, so I did not have anyone I could call and vent to. I went back to my office, gathered what I needed to take home with me, and left for the day. Home became my refuge. I had total peace there.

As I entered the classroom the next day, I did not know what to expect. I looked around, but I did not see the student. I was relieved. Ten minutes passed and there was no sign of him. I thought my words had been effective. Another thought was he was waiting to come and apologize before he came back to class. Three days later I received a memo from the registrar addressed to several other professors as well. The title of the memo was the name of the student. The registrar informed us that the student had been withdrawn from all classes that term because he was admitted to the hospital to undergo a psychiatric evaluation. I felt sorry for him. I wish I had known he was struggling so deeply. He did not know what he was saying or doing. At the same time, I was grateful he was not violent, and that it all ended well for his own good and the good of the class. However, I still could not reconcile the reaction from the department head. I had naively expected him to tell me to drop the student from my class. For this situation, I say naively because I did not know much about the American First Amendment. I am sure the department head based his advice on the right to freedom of speech.

This incident lingered in my mind for a long time. I started doubting my survival at UGA and whether I had the support I needed to make it. I started questioning myself about how I fit in with the campus culture and whether I was still naive about being accepted. I knew I was a strong-willed person, and when I put my mind to something, I can get it done. What I was not sure about was whether I had enough support from the department to be tenured. The third-year review was coming up and I had mixed feelings about my longevity at UGA. Then changes started happening in the Anthropology Department.

In the middle of my third year, the linguistics section of the Anthropology Department was uncoupled. The dean of the Franklin College of Arts and Sciences decided that linguistics would be a unit under the English Department. This was a result of a leadership change in Anthropology. A new department head had been hired and he did not see the added value of linguistics as a unit in the department despite the strong connection with anthropological linguistics. Before joining UGA, the new head was working at *National Geographic*. This scholar had secured a large external grant that UGA wanted him to bring to the university because of being hired. His interests were in ecology, which he quickly made a priority through new hires and new mandatory additions to the curriculum. Consequently, the linguistics teachers were asked to redesign their curriculum to include ecological linguistics. Many of the non-anthropology-based faculty found the mandate to convert theoretical linguistics, phonology, phonetics, and many other disciplines to reflect ecology a challenge. It was a long stretch. When it became clear that the requirement was not going to be met, the department initiated the uncoupling process. Consequently, the linguistics unit was moved to the English Department.

However, while some faculty members easily fit into that department, several of us were not a good fit. As a result, some left UGA for more traditional linguistics departments while others were placed in different departments like Classics, Computer Science-Artificial Intelligence, Romance

Languages, and Comparative Literature. I was the only one placed in Comparative Literature because of my attention to and focus on African languages, which would not have fit the English curriculum. Comparative Literature was and continues to be the most international department, offering world literature and world languages. Because of the non-Western language components, the dean decided that the African languages would be better served by that department. This was a wise decision. However, there were some difficulties. Based on my experiences in the years that followed, the challenges were enormous, and it was up to me to make it work.

The history of the Comparative Literature Department at UGA is complicated and interesting. Originally it was part of the English Department. These two related departments could not unify their interests despite their similarities, and in time, comparative literature was granted departmental status. When I joined the department, the offices were still under the same roof that housed and continues to house the English Department—Park Hall. My office used to be on the third floor, but only for a short time. When the department relocated to the Joseph Brown Building, just across from Park Hall, I received a new office on the ground floor. I was lucky to be on the ground floor because the building did not have an elevator like Park Hall did. Joseph Brown was one of the original campus buildings, and from what we have heard it was a girls' dormitory. In addition to its long history, there also is a legend associated with this building. Some say there is a ghost that appears at midnight. The ghost is that of a girl who killed herself in the dormitory. I did not worry about meeting this ghost because I am not a night person, so being caught late in that building was out of the question. I just hoped that the ghost did not linger around in the early hours of the morning since I was sometimes in the office as early as five both on weekdays and weekends, but I never saw a ghost or experienced anything unusual.

I liked my office in Joseph Brown because it had a lot of storage space for my language teaching materials, books, videos, and students' projects. The only disadvantage was its location near the bathrooms. The men's bathroom was right in front of my office door while the women's restroom was down the hall. I checked both and they looked similar without special additions. Then I got an idea. I pulled the signs off, typed up paper ones, and taped them on the doors. I designated the bathroom near me "Women" and put "Men" down the hall. Because renovations were still going on, I was sure the paper signs would be replaced with actual plaques. I told my friends this story and they thought I was crazy, but my wish came true as the signs were replaced as I had named them. Nobody knew I was responsible for the switch.

Establishing myself in a new department was not easy. I still taught linguistics courses, but I also put much of my energy into building the Swahili program. To ensure its sustainability in the new environment, I had to develop my grant-writing skills. I wanted to be self-sufficient in keeping the language sustainable, including hiring assistants to grow the program. All this was possible within two years of my move to the Comparative Literature Department because of my first grant from the US Department of Education. The grant was specifically for materials development, a project that allowed me to travel to Tanzania with a film crew from UGA to collect video footage for online teaching. This grant was instrumental because it paved the way for subsequent successes in grant writing and brought over $1 million in grants to UGA.

My work and success in writing grants paved the way to tenure and promotion in 1995. The financial security and the freedom to work on the programs that excited me were the main reasons I declined a position in 1995 from the University of Pennsylvania's Linguistics Department. While at UGA, I had trained their teachers for a year and provided support to their young African Studies Center by expanding the African languages program. When I got Pennsylvania's job offer, I agreed to be a visiting scholar for a year before I made my decision. But at the end of that year, I returned to Athens because I had achieved more at UGA.

— 16 —
THE AFRICAN STUDIES INSTITUTE AT THE UNIVERSITY OF GEORGIA

There was a single reason behind my eagerness to spend a year at the University of Pennsylvania (1995) and earlier, to interview at Yale (1992). After leaving Stanford, I kept looking for an African Studies program that was comparable to what I had experienced in California. Both Yale and UPenn fell short of what I was looking for. I did not think there was room for me to grow at either institution. Furthermore, I would be losing a lot of the independence I enjoyed at UGA. At that time, I oversaw the teaching of African languages, and I had latitude in how to organize it. I also had a lot of support from the department and the college. There was a sense of security for the program because its funding was attached to the department's funds. At Yale and UPenn, the funding for the program was from African Studies, given to the respective department that supervised the teaching of the languages. The source of this funding was not institutional but soft money (nonregulated money, which is given, in this case by the government, for a specific purpose and duration) from the US Department of Education Title VI funding. Thus, the sustainability of the language programs depended on the ability to continually receive the Title VI funding, a very rigorous triennial grant application competition. It was comforting to know that at UGA the African Languages program was secure if it continued to maintain good enrollment. I had set it up with a "can do" attitude, and I was determined to ensure success. For this reason, I was not willing to give up the program for an unknown future.

My anxiety about the UGA African Studies program came from the many ambiguities I was not aware of before joining the university. When I came for my first visit, the itinerary did not include any activities with the African Studies program. This was not an oversight. The African Studies program was still in its formative stages and only existed, at the time, as an organized interest group of faculty members with sharp instincts. However, I heard a lot about the potential for this program from Dr. Ben Blount and the Africanist faculty in the Anthropology Department. Ben was one of the co-founders and the first chair of this group's steering committee. Because the African Studies unit was built based on the membership of interested parties rather than as an institutional unit, departmental affiliations were nonexistent. To begin with, not all the members affiliated with the program enjoyed support from their departments, thus creating a dilemma if they wanted to foster collaboration or cooperation with members outside their own departments. It was also difficult to justify any involvement or activity done for or on behalf of African studies. The resistance from departments to claim credit for activities performed outside their departments and which had no direct bearing to their primary departmental assignments continued for many years, even after the group successfully launched an institutionalized program. Thus, associating with the African Studies program was counted as extracurricular, or at best, public service to the university. Interestingly, those who chose to stick with the African Studies program through thick and thin remained enthusiastic in the programs and activities that promoted the study of Africa on campus. Members organized both academic activities like public lectures and seminars about different aspects of Africa and, occasionally, social gatherings to strengthen collegiality and a sense of belonging. Members with research interests that focused on Africa were instrumental in promoting the study of Africa during the formative years. Their passion in promoting the study of Africa came from either experience in Africa as Peace Corps volunteers, time spent in Africa doing research for their master's and/or doctorate degrees, or a scholarly background from an institution that had a strong African Studies programs or schools funded by Title VI, which was established by the National Defense Education Act (NDEA).

NDEA was enacted on September 2, 1958, by the eighty-fifth US Congress and became a law the same year. NDEA was among many science initiatives implemented by President Dwight D. Eisenhower to increase the technological sophistication and power of the United States alongside, for instance, the Defense Advanced Research Projects Agency (DARPA) and the National Aeronautics and Space Administration (NASA). This law propelled to the top the history of the study of Africa in the United States because this is when global studies was established at institutions of higher education. The main purpose was to meet the needs of the United States government to protect its interests abroad, particularly in the areas of defense and national security because of the Cold War between the United States and Russia. Although the Cold War started in 1946, it reached its peak in 1957, a period also marked by the race to space exploration by both nations. This law was the impetus for the establishment of Title VI Centers across the nation. NDEA paved the way for the initiation of the centers whose funding was channeled through the US Department of Education, the basis for the development and sustainability of the teaching of African languages and area studies programs. The development included the continuation and expansion of Foreign Languages and Area Studies (FLAS) and Programs in African Languages (PAL), which, in the 1960s and '80s, were accelerated by pivotal forces resulting from Title VI funding. This is the background from which the UGA African Studies cohort emerged. The enthusiasm across different faculty programs and departments was to make UGA, though new in the game, a strong center that could compete with established programs across the nation. Many of us felt we had a good chance of succeeding, especially after establishing the African Languages program, a major element for a reputable African Studies curriculum at any campus.

Attending many African Studies meetings and social gatherings in my first year at UGA allowed me the opportunity to learn more about the program. I quickly realized that the setup was very loose compared to the programs I had seen at UCLA and Stanford. The chair of the steering committee was the de-facto organizer of meetings and activities that were considered of mutual interest to the group. Most of the activities involved organizing public lectures that depended heavily on volunteers from around the campus.

Once institutionalized as a program, it received $2,000 a year from the Dean of Arts and Sciences' special budget. This did not help to grow the program. In fact, on-campus lectures still depended on UGA African scholars, other scholars who were passing through town, or Africanists from sister institutions in Georgia. None of these lectures had an honorarium, offered travel cost reimbursements, or provided accommodations.

To succeed, it was inevitable that the group had to self-finance the activities through individual member contributions. Support came from the International Development Office, directed by the late Dr. Darl Snyder. He was an avid advocate of international links, a source of energy that led the UGA president, Dr. Charles Knapp, to be the first UGA leader to visit an African country, Burkina Faso, to sign institutional cooperation with the University of Ouagadougou. When Dr. Snyder retired in 1992, the African Studies program named a lecture series in his name. Up to his death, Dr. Snyder provided monetary support to the African Studies programs through an endowment established in his family's name.

In addition to building the African languages program at UGA, I remained a key figure in the African Studies program. There was little we could do because of limited funds, but we managed to transform the program from a resource center to an academic program that offered a certificate in African Studies. Starting the certificate program was a hard battle that took the group two years of back-and-forth discussions between participating departments, the college, the university council, and finally the Board of Regents. The process was started in 1991 by Dr. Salikoko Mufwene, who was the chair of the steering committee at that time. He left for Chicago after linguistics was de-coupled from anthropology and its faculty dispersed to different departments on campus. After Dr. Mufwene's departure, Dr. Ikubolaje Logan, a professor in the Department of Geography, assumed the leadership role and continued the process of adding a certificate program. It wasn't until 1993 that the program was finally approved.

One of the first things I learned, and that surprised me, was the program's strange modus operandi. The program did not have a designated home but resided in the steering committee chair's department and his/her office served as the designated program location. This quagmire continued for several years until I was appointed director in 1997. Somehow my predecessors were able to gain departmental sympathy that allowed the business manager to be the custodian and reporter of the allocated $2,000 program budget. The steering committee chairs also took the liberty of using departmental resources like stationary, telephones, fax and copying machines, filing folders, and other necessary program materials. Inheriting this disorganized structure in 1997 made me uneasy because I did not think I could, in good conscience, continue to lead our program in this way. I felt like the use of these supplies was stealing unless I was able to use part of the dean's allocation of $2,000 to reimburse the department for some of the supplies and other services rendered. I discussed this with the business manager of my department when the funds were transferred to her. She informed me that I could not reimburse the department because of the way the management of the funds was set up at UGA. She noted that, businesswise, it would be extremely difficult for her to account for the need to "JV" (a business term she used) meaning create a journal voucher, my spending account. In other words, she did not see how she could explain sending me a bill for office supplies to be paid from that account because we were not set up for that kind of transaction. She also noted that the money was earmarked for invited lecturers and travel to African Studies meetings that benefited the program. Then I asked, "What am I going to do?"

"Just leave it alone, it will not kill us," she answered.

This response did not ease my conscience and I had to think hard to find a sustainable solution to the dilemma. My priority was to find an office and to become independent from the departmental structure, which was not logical. That realization did not escape me for long.

One of the life lessons I learned when I arrived in the United States in 1981, was that one must create sustainable relationships, bonds, and connections. It is not easy, and sometimes such efforts might not yield friendships at the end of the day but might be the source of a sympathetic ear. In California as a graduate student, I developed two types of networks. The first I found in my classmates and colleagues; the second was with individuals in the community. I was successful at both UCLA and Stanford where I had a productive life, both academically and socially. The balance kept me sane with never a dull moment, even when I thought I had hit a dead end with a problem I was trying to solve. My friends used to say, "We are only a phone call away." I took them literally and felt I belonged to a village that gave me the same feelings I would have at my ancestral village in Tanzania. Many of my California friends have remained close to the point where I consider them family and, therefore, true members of my village.

Finding good friends did not change considering my relocation to Georgia. Such friends have always been there when I needed help. That is how I was able to resolve the African Studies space dilemma. One of the friends I made when I was moved to the Department of Comparative Literature (CMLT) was Dr. Betty Jean Craige, who extended a warm welcome by visiting with me at the old department before I made the transition to CMLT. She knew what I was getting myself into but wanted to assure me that she would be there if I needed help. She was also the faculty member the dean had mentioned when assigning my move to Comparative Literature. He designated her as my mentor. As a young assistant professor, I needed someone like her, and I owe a great deal of my success to her advice and guidance. She was the one I went to for advice on where to find space for the unit that had just been dumped in my lap, with a lot of expectations from the dean on its long-term success. When I became the director of African Studies in 1997, Betty Jean had just assumed an interim position as head of the Comparative Literature Department. That position gave her two offices in the building—her academic office and the department head's office. I was surprised that she offered her academic office to me as a temporary location for African Studies until we were able to figure out something else. She also promised to talk to someone on my behalf to find a more permanent location.

With the office space resolved, I turned my attention to how to set it up. I wanted to use this space to build a foundation for autonomy, independence, and, most of all, visibility. I sought advice from the Comparative Literature business manager, who gave me a tip on where to find free equipment. This was another pleasant revelation. The university collected all unwanted furniture and equipment into what it called the "Surplus Department." I visited this space and was surprised by the treasure. I was able to secure a printer, computer, and photocopier. I did not know how I was going to maintain this equipment, but I made up my mind I would use part of the $2,000 allocated for travel. I had to request permission from the dean to change the allocations and once I got clearance, I was determined to make things work. I did not need any furniture because the temporary office had a desk, chair, and shelf. I had an extra small table at home that I decided to take to the office to hold the printer. Bingo! By then, I had a respectable office we could temporarily refer to as African Studies.

My second task was to create a profile to continue building the program. I thought the easiest way to start was to establish a name through a letterhead and email address. I knew I did not have much money, but that did not seem to stop me from dreaming. I ordered letterheads for the entire fall semester. I knew that we would have more credibility across other departments on campus as well as with grant agencies if we communicated using our own identity rather than that of my home department. I found out later I was right.

However, I was unsure about who to ask or speak to about program support. I spent much of my time during the fall semester visiting different departments that housed faculty affiliated with African Studies. The goal was to introduce myself and to let the department heads and college deans know about my five-year plan as well as my short-term goals, all of which focused on the integration of African Studies in UGA's general curriculum. I received a warm welcome and an attentive ear at every visit. This was both encouraging and invigorating. The College of Education surprised me by offering a shared position with the college to secure a graduate student to teach Zulu (one of the major languages in South Africa). I was excited about this opportunity but worried at the same time because I did not have a guarantee for the other half of the funding that was expected from the dean of my college—Arts and Sciences. Because the dean of the College of Education made this offer, I persuaded him to speak with my dean about it. I was thrilled when he said he knew the dean and would give him a call. To my surprise, two weeks later, I received an email from the same dean confirming he had had a good talk with my dean and that they came to an agreement to co-fund a position for a graduate student to teach an African language. He asked me to contact my dean and proceed with the recruitment process for a suitable student to study education and teach Zulu.

To score a home run so early in the game was both invigorating and motivating. The graduate student recruitment did not take long because I had colleagues at different institutions who could provide me with a list of candidates. In a short time, the process was completed, and in fall 1998, we were able to start Zulu, a third African language, at the University of Georgia.

My focus was to build on the teaching of languages, which was the backbone of a successful African Studies program. I knew that with Swahili, which was started in 1988, Yoruba in 1996, and Zulu in 1998, the UGA African Studies program would be a national power to contend with. We needed a strong language program to increase our viability for external grants. The US Department of Education, for example, looked favorably upon institutions with good area studies track records and strong language programs. UGA was slowly building that credibility, and I knew I had a good chance to obtain these external grants.

In 1998, I started putting together a new grant proposal, and to my surprise, it was successful. I was thrilled that Friday afternoon when I received a call from the US Department of Education program officer informing me that I was one of the few selected as grantees for that cycle. Luckily, I was at home and not the office because I screamed and ran around the house like a puppy. It was a good thing that the windows were closed otherwise I might have scared my neighbors. For me, this grant was pure luck and I had hit two birds with one stone.

I was surprised when I got this grant because it had taken two rounds of submission after my last grant in 1993 for a study abroad program in Tanzania. The 1999-2002 award was my first major external grant after assuming directorship of the African Studies program in 1997. I felt encouraged and highly motivated, which gave me new confidence to forge ahead with plans to develop a non-Title VI African Studies program that had most of its support from the home institution. I wanted to prove that a successful African Studies program is not necessarily a Title VI Center (Title IV centers are administered by the US Department of Education and given to institutions to develop educational programs as well as provide financial aid to students), which was widely believed by many institutions.

While the objective of this three-year grant was to strengthen the African languages program by specifically developing electronic teaching materials, having an African Studies-sponsored grant was advantageous because the program received a small amount of overhead funds that could be used for critical programming needs, such as assistantships and office supplies. The bulk of the money, over $2,000 was earmarked for the development of online teaching materials.

Having worked with the first grant in 1993, I was confident that this grant would be easier to manage. The challenge with the 1993 grant was my lack of experience as well as the department's management of external grants. I was the first leader to bring in a large external grant to be managed by the Department of Comparative Literature. When the news came that I had received the grant in 1993, I felt like I was running scared because even though I had worked hard to put together a sound budget and justified the need in my narrative, I had no idea how I was going to manage the grant itself to meet the required goals. Since I did not have anyone to ask, I began to doubt myself and my success. I suddenly realized that I needed a professional cinematographer, one who would be willing to spend a month in Tanzania, running around the entire country filming a series of events. Then, there was the editing of the footage to create the needed vignettes that would make the twenty-three video lessons I had designed. The only trick I had in my back pocket was the experience from a mini video project I had started while at Stanford and finished during my first two years at UGA. I knew how hard it was, especially when one also had to balance teaching and other responsibilities. I needed cinematography and video editing pros. I suddenly realized that this part of the planning should have been done before submitting the proposal. But the narrative was convincing enough that the reviewers did not raise questions on the production aspects of the project.

After a few days of worrying, I decided to approach the director of the Center for Teaching and Learning and offered to pay the center to take charge of the production portion of the project. I knew the staff at this unit since I had worked with them on my mini video project. My fears were eased when the director agreed to assign a cinematographer and pledged to have a team to work with me in the editing process and the final stages of creating the different segments for each lesson. Once that collaboration was secured, I was sure my project would have the integrity I had promised in my application.

Putting together the 1999-2002 grant, I was cognizant of my weaknesses in the 1993 project. I clarified my concepts, goals, and objectives. While writing the proposal, I secured collaborators from UPenn and Indiana University. With that team, coupled with the office staff at the Center for Teaching and Learning, I knew I was developing a hassle-free project.

The grant raised the African Studies profile immediately. The dean allowed us to change the name from "Program" to "Center." The argument was that as a "Center," we would gain national credibility and be more competitive with other "Centers" for federal grants. The grant supported our argument and convinced the dean that we needed the name change. This success made it possible for the college to increase our annual budget from $2,000 to $10,000. This bigger budget allowed us to do more, and I was able, for the first time, to hire additional staff. Having an office manager and an administrator removed the pressure of running the unit single-handedly while teaching four classes, managing two grants, directing the African languages, and running a study abroad program in Tanzania. I realized new hope and enthusiasm and was determined to do more with less. Suddenly, my goal to get national and international recognition for the African Studies program at UGA seemed realistic. I finally felt that my decision to stay at UGA was the right one. There was a possibility to build what I had envisioned, a prosperous African Studies program on campus. I pushed myself harder and went to the office seven days a week to make sure the momentum was not lost. Although I did not think about it then, the hard work did pay off in the years that followed. The harder we worked, the more successful we became.

The success of the first two helped me perfect my grant-writing skills. I was encouraged to put a proposal in each year, and, to my surprise, I received a grant each time. The 2001 grant for undergraduate studies curriculum development was well-received by my home department, the college, and the central administration. For me, it was another motivating factor to stay on target. My goals seemed doable, considering the success achieved in just three years of my directorship and two grants back-to-back. With the two big grants, I knew I had a bigger cushion, which was money I needed to add to programming. But I felt we could take another leap. This time it was for Yoruba language teaching materials development. Instead of developing a collection of video materials, we proposed gathering online materials, like we did for the second Swahili project. My collaborator for this project was the Yoruba language instructor, who was a member of the African Studies academic staff. In early 2004, we were notified that the proposal was successful. UGA was now very well known by the US Department of Education for successful grant administration through African Studies. It also was a victory that we had achieved "institute" status, which was granted in 2001 by the Board of Regents.

The next big grant project, initiated in 2003, was even more ambitious. As a unit, we wanted to make UGA one of the leaders in African Studies in the Southeast, but most of all, the leading institution in the state of Georgia. To do so, I had an idea for a proposal that would collaboratively unite all the institutions that had departments that supported the study of Africa. Most of these departments were history, geography, music, and religion. To some extent, many of these institutions had faculty with interest in the study of Africa but did not have a structured organization akin to what UGA had in the early years of forming an African Studies program. This provided an opportunity for a proposal to use UGA's leadership to create a statewide focus for the study of Africa. The grant was intended to provide funding for the development of African languages at these institutions in addition to courses that had between 50 percent and 100 percent African content. To do so, institutions had an opportunity to make additional hires and to receive funding for African Studies faculty research.

To say this was an ambitious project is an understatement. I was naive about how hard it was going to be to work across institutional structures and bureaucracies, especially where grant writing, administration, and policies varied so much. At this time, I had a valuable development and grants committee in place. We selected three institutions, Fort Valley, Savannah State, and Georgia Southern universities for this collaborative project. The idea was that once the objectives were met, namely at the completion of the grant and implementation periods, the four participating institutions would mentor other state institutions until all of them had an African Studies program on their campuses. It was not easy for faculty at the selected institutions and key administrators to agree on the objectives or the most basic elements of the grant, such as cost sharing. There were many discussions throughout the proposal writing and many visits by key UGA players to these institutions. Although the application process was tedious, we managed to complete it on time. We were not optimistic because the budget was over $400,000 and involved so many moving parts. To our surprise, a phone call in early 2004 changed our pessimism into a celebration. The grant was awarded to UGA for over three years.

After this, and with multiple concurrent grants to manage, it was wise to slow down and pick smaller projects. In subsequent years, I embarked on less-ambitious grant applications and focused on students' international experience through UGA's study abroad program. I was a veteran in this area having successfully obtained a study abroad grant from the US Department of Education in 1993. I had maintained this program over the years in addition to the major grants that targeted the development of the institute. I used my success in developing the institute to convince the US Department of Education I could run multiple funded programs. Consequently, I received study abroad funding from Washington every three years, from 2002 through 2008. In 2008, when I was contemplating stepping down from the directorship of the institute, I decided to initiate an international study abroad experience for K-12 teachers to bring that community to the UGA campus. That proposal was successful too. We did not start the program in 2008 when we received the funding because I needed time to plan it well. The Department of Education granted an extension on the start time, and, in 2009, the first group of K-12 teachers went on a study-abroad tour in Tanzania. Although I could have restricted it to just Georgians, I became a little ambitious and advertised the program to the entire nation. Doing a pre-trip orientation was challenging, but I got help from the Center for Teaching and Learning at UGA. I was assigned a technician who created a virtual meeting with students across the nation. I would like to think it was Zoom before Zoom. There were a lot of glitches, but it served our purpose.

 The first program was very successful and created a lot of excitement in the K-12 community of teachers. This pushed me to dare to request funding for a second group. The reviews and participants' testimonials made it possible to secure a grant a second time. After that, the program became very popular and received biannual funding allowing it to continue. Because we had a lot of applications from Georgia, I decided to recruit more heavily there and occasionally include out-of-state participants. When I retired in 2018, the program was transferred to the UGA College of Education and the focus was extended to include science and technology, particularly robotics. The program is still growing strong with funding from the US Department of Education. In 2022, the program received an additional grant because of the two-year lapse due to COVID-19, allowing more participants in the program instead of the usual twelve.

— 17 —
Growing the African Studies Program

In 1999, when the African Studies program became the African Studies Center, we moved from the temporary space to a two-room office assigned to us in Candler Hall. This was both necessary and inevitable. The Comparative Literature Department underwent a drastic restructuring, and the new administration did not like the idea of housing a unit that had nothing to do with the department. There was a lot of tension in the department, creating a toxic and confrontational environment. There were rumors that the new department head had been told who had not voted for her hire, and these individuals became outright enemies. Included in the list of the undesirables were friends of those who had been placed on the enemies list. Consequently, the department had two camps, and I was in the so-called enemies' list camp. This was the beginning of a toxic atmosphere that lasted for three years. The effects included my promotion to full professor being delayed until 2007 because the department head refused to put me up for promotion.

Losing the donated space for African Studies was both a blessing and a curse. A blessing because the department's leadership had created a hostile working environment, and a curse because I had no other space to operate from. The initial problem was what to do with the equipment I had acquired and the accumulated program files I had on the shelves. My academic office was too small to accommodate the computer, printer, and boxes of files. Furthermore, I needed special permission from the department head to use my office for activities related to African Studies. Suddenly I had a mountain of problems with the new administration, and therefore, the wildest time of my tenure at UGA.

Meanwhile, I decided to focus my energy on finding a new home for the African Studies program. I approached the dean and requested more space, explaining the circumstances. This was an opportunity to inform him about the program and its impact campuswide. After a long discussion with the dean, he asked for a mission statement. Luckily, I had prepared one and he officially designated me as director. I presented the dean with a list of program needs. He agreed that we should change African Studies from program to center to increase its national and international visibility. With this development, the dean supported my request for new space and relocated African Studies to another building, Candler Hall. He also made a provision for me to hire an academic professional as an administrator. I was able to hire Dr. Leonce Rushubirwa who was finishing his Ph.D. at the University of Alberta Canada. Finally, this stressful situation had a happy ending.

The two-room office space in Candler Hall was a big and welcome promotion for a unit that had never been independent. Having an administrator was also an achievement because it was no longer a unit that was run by one person. The new administrator assumed some of my responsibilities so I could have time outside my scheduled teaching, grants management, directing the African languages program, and leading a study-abroad program to focus on critical restructuring plans and the overall development of the African Studies program.

We shared the allocated space with the Institute of African American Studies whose director was Professor Ronald Miller. The institute had more space allocated to it, but administratively, like African Studies, it had a limited workforce, just a director and only one administrative secretary. However, unlike African Studies, the institute had a large pool of jointly appointed faculty (that is, each faculty position in African American Studies was shared with another department, which was considered the home department and the tenure- and promotion-granting unit). The institute controlled the faculty members' workload, requiring them to teach a certain number of African American Studies courses. African Studies had not reached that level of consideration from the university. African Studies-associated faculty could only keep their interest in Africa and affiliations with the center by developing a course with African content and seeking approval from their departments to teach it. Thus, we did not have the ability our neighbors had to teach and organize activities of interest to faculty and students. Most of our operations and activities thrived on volunteerism.

Because the Institute of African American Studies had more space, including four offices and a conference room that doubled as library space, the director invited us to share the conference room/library space, allocating two shelves for our books. The "neighbor" relationship soon turned into a "big brother" relationship. The director had a larger budget, which was funded at the level of a department and had additional resources from endowments, and he was able to get new furniture and equipment each year. Instead of disposing of what the institute did not need, he passed it on to us, mostly furniture, computers, printers, and other office supplies. We were happy to receive them. By now we were used to hand-me-downs because everything the African Studies had at this point had been acquired from the surplus store on campus.

We only stayed in Candler Hall for a year before we were notified that the space had asbestos that had to be remediated. Our new location was in the Academic Building, and during this time, we had a new provost, Dr. Karen Holbrook As we were preparing to move to the Academic Building, I spoke with her about our space needs. Instead of being joined at the hip with the African American Studies Institute, this time we were considered individually and assigned a larger space with four offices, a breakroom/utility room, library, and conference room. This was supposed to be a temporary location. The administration had promised the unit a stand-alone building on Waddell Street, placing us at the same level of consideration as the Latin American and Caribbean Studies Institute (LACSI). With that possibility, the dean's office did not see the need to provide funds for furniture or new computers until after the move to a permanent location. The space also was not fully renovated. We were given floor maps for our permanent location and asked to provide a remodeling wish list. This was exciting.

The African American Studies Institute's space in the Academic Building was permanent, so it got refurbished, and they donated all their old equipment, which was in mint condition, to us. We accepted gratefully while anticipating new equipment as soon as our permanent location was remodeled. Little did we know that this was not a solid promise. We waited two years and, each time we inquired, we were told there were other pressing needs on campus but that we were still on the priority list. With the provost still on campus, I had hoped that the promise would be kept. In 2002, she left UGA to become the president of another university. Her departure changed the promise of a house on Waddell Street and eliminated us from the list of consideration for remodeling, new furniture, and equipment in the current space. In addition to the provost's departure, we had another setback that year, the departure of our administrator, Dr. Leonce Rushubirwa, to Louisiana State University, where he had secured a teaching job.

These setbacks did not dampen our enthusiasm. We knew success would not come easy, but it was worth aiming for. With the grants we had and those we were working on, we were convinced that the catalyst for success was the ability to secure more external grants.

The new grants afforded the center more money to hire a full-time administrative secretary. When we announced the position, we attracted a lot of applicants. But we chose Loretta Davenport who had a similar role at the African American Studies Institute This concerned me when she applied because of our relationship with the institute. Contrary to my fears, the relationship between the two entities was not affected by Loretta's move to African Studies. Ron, the African American Studies Institute director, found a replacement right away.

There was no need for a transition period when Loretta moved to our office. We had known each other from Candler where we had shared space and she had trained our administrator during his transition period. She had looked out for us, letting us know whenever her unit was discarding items and saving the best for us and helping set up the computers and printers.

With her experience as an administrative secretary, Loretta taught me a lot about the way UGA operated. Watching how hard she toiled made me stop pitying myself because of my workload. She made me aware that many women on campus covered multiple assignments. She had two jobs rolled into one—office manager and administrative secretary—for one administrative secretary salary under the Institute of African American Studies. By moving to the African Studies Center, she was able to earn a bit more money because I convinced the dean to retain the funding allocated for our outgoing administrator's salary, arguing that the pay was set based on the position and did not take into consideration the individual's academic achievements.

Shortly after Loretta migrated to African Studies, we received another major grant for the development of the Yoruba language at UGA. Hiring Loretta strengthened the African Studies program by providing a stable administrative arm during a critical time. Grants management requires diligence and expertise in keeping up with the agencies' requirements. Loretta realized I was already spread too thin, and I was struggling to create a sustainable program by securing external funding to leverage the support we were getting from the college. Loretta's support and professionalism during this time was significant and enhanced the center's progress.

At the beginning of 2003, we received another call for a proposal. This one was interesting because it combined language and curriculum development. I saw this as an opportunity to support both the languages and the certificate in African studies. Even though we had just received a second grant from the same agency, I did not feel it would be presumptuous to try for this one too because I had a good track record in grants management. I also saw an opportunity to convince the college that the unit needed more than one assistant, specifically an office manager. My gamble paid off, and we received both the dean's approval to hire an office manager in 2003 and the grant in 2004.

For the office manager's position, I set out to find a creative and innovative individual with managerial experience. I approved the hiring of Amsale Abegaz who was an office manager at the Office of Minority Services. I had met her several times during the early years of my directorship. Her boss was a very progressive individual who welcomed collaboration on cultural projects. The African Studies Center and Minority Services sponsored many cultural community programs including Africa Weddings, Taste of Africa, and cultural awareness activities. All these programs were extremely successful because of Amsale's creativity, professionalism, diligence, attention to detail, ingenuity, deep interest, and devotion to international cultures. After her boss left the university for another institution, all collaborative endeavors ceased. When the opportunity to hire came up, I decided to be bold and approach her about the possibility of taking the position. I was convinced that her love for programs on Africa would spur her to accept. She took a week to think about the offer, which seemed like an eternity. That is how desperate I was to win her over. At the end of a week, she came unannounced to the office in the Academic Building. We talked for a while about unrelated issues, and then she looked me straight in the eye and said, "I have decided to take the position." I did not hide my relief.

I stood up and hugged her, thanking her repeatedly. Because the advertisement had already been posted, I urged her to apply as soon as possible. As I had suspected, she was the most qualified person for the position and had the most experience. We received ten applications and no one else who applied came remotely close to her in experience and achievement.

Running an office that had morphed from one person, which was me as the director, to a full-time staff of three for administrative duties was a dream come true. Once Amsale joined us, I realized how lucky I was to have these two women working with me side by side. They had a passion for what they did and saw something good coming out of our hard work. Their combined experience and diligence inspired me to do more and to dream bigger. Amsale made me realize that positive thinking was the most important factor to achieve what we were doing, and we should look at setbacks not as problems but rather, opportunities. Nothing that came to us was too big to handle.

With the support of these two women, we embarked on multiple projects, including the most significant one—the application to become an institute. Amsale joined us at the height of the African Studies grant application success with cumulative awards of nearly $1 million. With such significant funding, we needed a business manager to handle accounting. At the time, 2004, we had received a grant of over $300,000 for the development of programs within the University System of Georgia to strengthen the study of Africa. For this project, I assembled a team of three grant-writing experts: Dr. Jack Houston from the College of Agriculture, Dr. Jennifer Frum from the Office of the Vice President for Public Service and Outreach, and Dr. Christa

Hoffacker, a temporary African Studies staff member. Putting the proposal together was a daunting task because of the amount of data needed from the participating USG institutions. It was simplified because of the support from the office staff, a rare advantage that we did not have in previous grant proposals.

We were ready to expand the center's activities to include study abroad in Africa, a viable possibility because I had both Amsale and Loretta at the African Studies office. They worked well together and went beyond the call of duty to get tasks done. Often, Amsale stayed in the office until 8 p.m. to finish or push important projects forward. One memorable event that she finessed was the tenth anniversary of the annual Darl Synder Lecture Series, which was the unit's flagship event. We had planned to invite renowned journalist and UGA alum, Charlayne Hunter-Gault, to arrive from South Africa to speak. Her credentials as a journalist included her tenure with National Public Broadcasting and Public Broadcasting Service news and CNN. She also was a recipient of the Peabody Award administered by the Henry W. Grady College of Journalism and Mass Communication at the University of Georgia. She was a celebrity on campus because she was the first Black student to enroll at UGA as an undergraduate during the South's segregated period. With this background, she drew a big audience from the university, the community, and the state. The university agreed to make Hunter-Gault's talk a "charter lecture," an honor not easily granted by the UGA president's office. This made me feel that the African Studies Institute had made its mark on the UGA campus. It was an important event, and I knew I could count on Amsale and Loretta, to make us proud. Without fail, they delivered beyond my wildest imagination. I was impressed by Amsale's extra hours each day in preparation for the event. She came to the office seven days a week for more than three months and stayed late each workday to make sure the plans were fully met. As a team, the two women displayed a level of creativity that made the event both successful and memorable.

Amsale continued to amaze me as she settled into her new role as business manager in addition to being office manager. The experience she brought transformed the institute from a unit that depended on hand-me-downs to one that was deserving of what other units took for granted. She could advocate for the unit better than I could because of her experience on campus and knowing how to seek out what we needed. She surprised me one summer when I came back from study abroad to find the entire office refurbished with new furniture, new computers, and a multimedia meeting room. I was so surprised that I said aloud, "Which bank did you rob?" She smiled and said, "I just asked for what we deserve." She was able to accomplish this because she had excellent public relations skills and was knowledgeable of how the system worked. Most of all, Amsale was excellent at managing people and funds. At one time, we were working on four grants simultaneously in addition to several study-abroad programs. This meant a great deal of accounting and reporting. If it were up to me, she would have received better pay than what she was getting. Unfortunately, what one was assigned to do at UGA did not always match their compensation, especially for individuals in specific areas of service. It is painful to even think about it, but that is the way things were and still are. Women seemed to suffer more in this system than men. During my leadership at UGA, I could not credit the success of office operations to anyone other than the African Studies staff and its two very dedicated women, Amsale Abegaz and Loretta Davenport, who worked tirelessly even though what they earned did not match the work they did. They knew it was not for lack of trying. I requested raises for them numerous times without much success. They stuck with me because they believed in the mission of building a strong African Studies program at the University of Georgia.

As the director of the African Studies unit, I realized from the onset that our mission was ambitious with a difficult, if not impossible, goal. In addition to having two wonderful women to help me run the unit, I needed the energy that had kept the study of Africa on the campus years before I came to UGA and became an active member. African Studies had achieved much over the years through the passionate efforts of its members. This translated to enthusiasm and volunteerism. The patrons' primary responsibility and loyalty was to their home departments, anything else was over and above their departmental assignments. Volunteering posed some problems for some members in their departments, especially if they were not yet tenured. Taking up duties outside their departments was interpreted as a time management issue, diverting research time to activities that did not count. As such, some members did not disclose their association with the African Studies program and did not want me to send annual letters to their departments outlining their services

To accomplish what I considered essential in building a credible African Studies program, I needed to develop interactivity campuswide, which required time and commitment from its members. It was not easy to accomplish that through volunteerism, but I was determined to keep in mind past experiences and find a way to make it work. I decided to create committees and approached different individuals to request their participation. I needed several committees: an advisory board, development, curriculum, and special events and conferences. I needed at least three people for each committee, a total of fifteen. I sought individuals who already were accomplished, either professors or associate professors. I was elated when they agreed to serve on these committees. We wanted to create a permanent spot for Africa on the UGA map. I knew I did not own or reward the volunteers for their time, but their diligence paid off in the success African Studies enjoyed between 1998 and 2010.

Once the committees were set, the restructuring work began. The toughest part was making sure these committees were productive and that we had something to show by the end of a semester or the overall year. I hosted many committee meetings and was very lucky that all members showed up. This was a clear display of commitment and eagerness to bring to fruition what we had dreamed about and planned. All the committees were very generous with their time. I was careful in choosing meeting times, mostly during lunch when faculty could leave their desks. I also paid for the lunch, though I never told them this. I did not believe in using institutional money, even though we had discretionary funds. I used those funds for the annual open house that supported minority and struggling students who wanted to participate in study abroad in Africa. I also used it to cover the meals for administrators and invited guests who we felt should not be asked to pay.

The development committee was also a de facto advisory committee. It was charged with creating a five-year plan and conceptualizing funding strategies. Members of this panel met more often compared to the other committees, at least twice a month, often weekly. I also scheduled extra meetings with experienced grant writers. I would like to pay special recognition to Dr. Jack Houston on this front. I was in his department so often that his secretary and a colleague in the College of Agriculture thought I should get a permanent desk to reduce my campus crisscrossing. Dr. Houston was known as the "math guy" and focused on shaping budgets for the different grants and generating the narrative for the budget justification. He claimed that I went to his office often to keep him on his toes on this major task. To tell the truth, when I had a task on hand, I pushed hard. Sometimes too hard. If all these people who worked with me did not love me and believe in the mission, they would have built a bunker and hid there to avoid me. But they did not, and I am eternally thankful. Jack's office neighbor used to shout out to Jack whenever he saw me coming, "Here she comes." All Jack did was laugh as I walked into his office, warmly welcoming me. He would stop whatever he was doing to hear me out. I was like a hurricane that kept coming without a warning.

The hallmark of this committee, including our superheroes, Amsale and Loretta, was its success in following the money. We raised a lot of cash for different projects and activities, and we were able to attract several major external grants that allowed us to expand programming, hire staff, and acquire equipment and furniture. Finally, we could afford new items and no longer needed to search for used goods in the university's surplus store or other departments. Amsale had exceptionally good taste, and she made the office warm and welcoming. She also decorated our space with pictures of past activities and annual events to show visitors our hard work and dedication.

The development committee was instrumental in building outreach to other institutions around Georgia using the 2004 grant, particularly those that did not have a robust African Studies program. The objective was to select a few pioneer institutions from around the state that would embark on pilot projects to strengthen their curriculum focus on African Studies and African languages. UGA's role was to coordinate the proposed activities, which required periodic visits to these institutions. Because of our experience, UGA served as a role model for how to grow a visible presence of Africa on campus and how to develop a curriculum that could be incorporated into the base institutional curriculum.

The curriculum committee was another very engaged group. It was responsible for the establishment of a robust curriculum with a rich African Studies focus across disciplines. Their first task was to find existing courses at UGA with African content. With that list, I was able to approach department heads to request cross-listing privileges. It was a tough sell considering territorial integrity, but we were modestly successful. Some departments introduced restrictions on the number of students that could be allowed to cross-over. The biggest successes came from departments with active Africanist scholars such as agricultural economics, anthropology, history, comparative literature, geography, international agriculture, education, forestry resources, family and consumer science, social work, and religion. This major support allowed for a surge in enrollment numbers in the capstone course, "Introduction to Africa," whose numbers allowed a transition from a single offering to two sections of over fifty students a year. Later, the number grew to over a hundred students in each section, a total of four hundred students in a year.

The special events committee was a critical one because it was responsible for inviting occasional and annual lecturers. We had two main annual events, the fall lecture, and the annual Darl Snyder Lecture. For the fall lecture, the Institute of African Studies targeted prominent leaders from private agencies and government services, along with scholars. They brought diverse views of Africa and attracted large audiences. Such lectures also were important for students, some of whom had limited knowledge and/or exposure to Africa. We were able to bring people from the World Bank, the Carter Center, the US State Department's Africa desk, along with academics, directors of well-established Africa Studies Centers, and renowned playwrights and literary critics. We made sure the disciplines were varied to cater to our diverse pool of academic interests on campus. The committee was instrumental in establishing a list and a selection criterion for prospective invitees.

The Snyder event was our signature lecture of the year. It occurred in the spring and was popular because of its namesake. Dr. Darl Snyder was a prominent and longtime UGA professor, having worked and directed the office of International Development for decades. He facilitated keen interest and support for the study of Africa on campus. He encouraged faculty to establish a disciplined focus and was a pioneer of the study of Africa at UGA. His support included expanding the number of students from the African continent who came to study at UGA. Many of these students were from the sciences, particularly poultry science and agriculture.

Dr. Snyder became affiliated with the University of Georgia in July 1969 as a program specialist at the Rural Development Center in Tifton. He became director in July

1972 and remained at RDC until 1974. In September 1975, Dr. Snyder was appointed as the director of international agriculture programs on the Athens campus. In 1977, he was appointed as the campuswide coordinator of international development. He then became director of the Office of International Development at UGA in 1989.

I looked up to Dr. Snyder as my role model because of his wisdom and tenacity. Throughout his career, he kept his eyes on the prize, and his determination enabled him to achieve much success. During his service, he initiated the development of a proposal that led USAID to award UGA a contract to implement a Collaborative Research Support Program on peanuts (Peanut CRSP) in 1980. To date, this program has brought UGA more than $35 million. Then, as technical director, Dr. Snyder played an instrumental role in UGA's partnership with Tuskegee University in the implementation of a USAID-funded agricultural human resources development project in Upper Volta (now Burkina Faso). One other quality I admired was his big heart, for he was always encouraging and supporting faculty in the development of an African Studies Program at UGA, which was approved and implemented in 1987, a year before I was hired. After he retired, he and his wife, Florence, did not stop their support and affiliation with African Studies but remained involved with the institute and various other international offices. He encouraged me throughout my tenure as the director of the African Studies Institute, especially when the road forward became tough. He always said, "The sun will always rise, and it will set, no matter what." He never missed a chance to cheer us on every time we reached a milestone or achieved noticeable success in our efforts to make African Studies at UGA the best in the country.

During his impressive twenty-seven-year career at UGA, Dr. Snyder made over thirty-five trips to Africa, establishing partnerships with governments' agriculture and education sectors to support societal development, particularly in Burkina Faso. His interest and ties to that country began during the first visit by a UGA president to Burkina Faso to sign an institutional cooperation memorandum with the University of Ouagadougou.

To honor his commitment to the study of Africa, the African Studies Institute established an annual lecture in Dr. Snyder's name upon his retirement in 1992, four years after I came to Georgia. The idea for a lecture series came up during a coffee meeting with my friend and colleague, a sociology professor. As a graduate of the University of Wisconsin, which had an active and successful Center for African Studies, my colleague had valuable ideas on how to achieve our goals. The lecture idea came from a discussion on how to create an event that could bring the campus to the institute through public lectures so that our mission to focus on the study of Africa would be understood across disciplines. We came up with the idea of honoring Dr. Snyder at his retirement by establishing an annual lecture in his name. We were convinced that his work at UGA was well recognized across campus specifically as director of the international development office and as a professor of international agriculture. He also was well-known across the state because he introduced blueberries to Georgia. The lecture series was the least we could do to thank Dr. Snyder for his generosity and contribution to the study of Africa at UGA. A foundation in his name also was started by my predecessor as director of African Studies, Dr. Johanne Buis, who received the first donation from the Snyders' family doctor, Dr. Farris T. Johnson.

Dr. Snyder enabled African Studies to open the doors to the outside world by financing many of its lecture activities. Unfortunately, Dr. Snyder died February 25, 2017, just eight days short of his ninety-fifth birthday, the week the African Studies Institute was celebrating the twenty-fifth anniversary of the lecture series. He had left his home in California, where he had been living, accompanied by his daughter, Dr. Cherie Snyder, to attend the celebrations. He fell in his hotel room on February 20, 2017, one day before the lecture, and died at St. Mary's Hospital. It was like he had come specifically to say good-bye and to make sure he died in his beloved and favorite city. He died among his friends and those who cherished what he and his family had done for UGA. His beloved daughter, Cherie, a UGA alumna with master's and doctorate degrees in psychology, was at the hospital with him. This was also the hospital where his wife worked as a speech therapist before she retired. I was fortunate to visit with him at the hospital the day he was admitted and talked with him before he drifted into a coma. My memory of him is that of a true gentleman and a scholar, a role model to many for his spirit of service, compassion for others, and commitment to lifelong learning and mentorship. He had an incredible panoply of cherished friends all over the world, all of whom will miss his quick wit, twinkling eyes, and indomitable spirit. After Dr. Snyder's death, Cherie decided to retire and relocate to Athens from California. She has remained my great friend and sister.

Besides the annual lectures, other activities undertaken by the curriculum committee included the summer study-abroad programs. The inaugural program was in 1998 when the University of Georgia Foundation Fellows traveled to

Tanzania for one month. The Foundation Fellowship is a top academic scholarship that provides generous academic funding, stipends for group and individual travel-study, and research and conference grants. The visit to Tanzania followed from an invitation from Ambassador Gertrude Mongella, a diplomate from Tanzania, during her visit to UGA in 1996 to give a lecture on human rights. This invitation was made at a dinner organized by the Foundation Fellows program in honor of Ambassador Mongella who was a guest speaker at a lecture organized by the UGA Humanity Center. The Humanity Center director, Dr. Betty Jean Craige, had heard of Mongella from her service as the U.N. Secretary General for the 1995 Beijing Landmark International Women's conference. Betty Jean invited Mongella to speak at one of the annual international lectures organized by the Humanity Center at UGA. The Foundation Fellows and the program leaders were thrilled by this invitation and started planning for it immediately.

I was not aware of the plans until fall of 1996 when I was invited to a lunch by the program coordinators, Dr. Kathleen Harris and Dr. Sandra Whitney. l listened intently to their plans. I suggested a pre-trip to Tanzania for logistics purposes. They agreed to fund me to travel to Tanzania during the spring break of 1997. After my trip I was able to offer both academic and cultural trip plans, which were accepted. The Foundation Fellows traveled to Tanzania in the summer of 1997, marking the inaugural UGA study abroad program in Africa.

The program was rich, both in academic and cultural events. The academic side included attending a session in Tanzania's High Court, marine science lectures and tour of the Zanzibar marine station, visits to the Tanzania Parliament and the office of the Dar es Salaam mayor, and a field study on ecotourism. The cultural events included many community activities, such as attending a two-day traditional wedding and climbing Mount Kilimanjaro.

The success of the inaugural trip to Tanzania sparked interest and opportunities to study abroad in different parts of Africa. In 2000, the institute supported trips to three locations: Tanzania, Kenya, and South Africa. In 2002, a fourth program emerged for Ghana. All these programs focused on the African Studies capstone course, "Introduction to Africa," with special focus on courses that included anthropology, geography, science education, natural history, ecology, agricultural economics, home economics, business administration, language-gender-culture studies, Swahili language and culture, social work, and public health. The Tanzania program remains the oldest and most successful program in Africa.

Without a doubt, the programs contributed to the growth of African Studies at UGA, enabling it to develop from a program to a center in 1999 and finally to an academic institute in 2001. Its visibility on campus also was instrumental in shining a light on other area studies at UGA. Other programs began to develop, like those run by the African Studies Institute. For many of the Africanists on campus, this was the goal; their endurance and hard work had paid off. The African Studies Institute allowed Africa to be a permanent feature in the curriculum. The institute's continued health depended and continues to depend on strong staffing at both the administrative and academic levels. Sustaining such a thriving institution requires consistent funding, mostly through institutional commitment as well as external grants for special research and development projects. The various grants, dating back to the late 1990s, kept the institute busy and visible on campus and beyond. Having additional funds allowed for restructuring as well as a much-needed transformation that made African Studies an integral part of the UGA curriculum.

Studying abroad was the most rewarding aspect of my engagements at UGA. Often there is not enough time during classes to get to know students and listen to their stories, learn about their families and friends, and develop a learning environment that is both challenging and rewarding. Studying abroad in Tanzania was considered nontraditional because, unlike the European, Asian, and Latin American programs, we did not have an academic campus where the students lived and attended classes. The Tanzania program was a combination of group learning and field experience. We emphasized learning by seeing and doing and not just by reading and lectures. It was always refreshing when, at the end of the three-week program, the students' evaluations indicated they had learned more than a year's worth of material. Furthermore, they retained the subject matter more while on the ground than in a traditional classroom.

Maymester in Tanzania was designed as a multidisciplinary program covering both science and humanities. Science professors traveled with the students to emphasize their disciplines and to dispel the myth that Africa was only good for language and culture. During the twenty-one-day program, students were allocated five days for community service. This allowed students to interact with the people and learn about their culture and daily activities. This became the favorite part of the program—students had an opportunity to be creative in finding sustainable solutions to endemic problems in the community. Two areas emerged, education and health. Students donated ten dollars for a project of their own design to help one early childhood school.

When we visited the school, we found the students sitting on the floor because they did not have desks or chairs, nor did they have anything to write on or with. They used their fingers to write in the sand, and the teacher walked around to grade their math and writing exercises. When we visited the following year, we were pleasantly surprised by what they had been able to do with ten dollars per student, or a total of two hundred and twenty dollars. It was enough to procure a shared desk, a chair for each student, some notebooks, and a locker to store the teaching materials. Additionally, students wore nice uniforms: the girls in plaid dresses and the boys in plaid shirts and khaki shorts. The school was near a church, and the parishioners were impressed by what we had done and donated supplemental funds to provide the students with hot cereal when they arrived at school. Many of the kids came to school hungry because their families could only afford one meal a day, which was the evening meal.

Our next project was at a primary school in Mwanza. This project was initiated and executed by four students who designed it after returning from studying abroad. We had visited Buzuruga Primary School where our students were given teaching assignments. The subject areas were geography, history/political studies, biology, mathematics, English, and music. Two of the students, Toddy Ferry and Phillip Adams taught a class of one-hundred pupils huddled together around ten shared long-bench desks. Their idea of a sustainable solution was to raise funds and collaborate with the community to build an extra classroom. They hoped that after the initial project, the community would be motivated to add one classroom each year. The four students were successful in raising $10,000 for the project. They returned to Tanzania and stayed with the community for one month while working with them to build the school. The funds were used to purchase the construction materials while the community provided the workforce. When the classroom was completed, they painted a mural on one of the outside walls depicting the UGA Arch superimposed over the map of Africa.

The success of these students' efforts became a major motivation for the community. They used it as the basis for a major grant from the World Bank, which funded twelve additional classrooms. Subsequently, another group of pre-med students developed a project focusing on the health of the community. After visiting a local public health center, the students wanted to do work to improve children's health, specifically on malaria. They raised enough money to buy pretreated mosquito nets that they distributed at three prenatal clinics in the community. They also put mosquito screens on all the windows in the children's ward at the hospital. Before they traveled to execute this project, the students sought donated laptops that they refurbished and equipped with programs that would help doctors access material to advance their professional knowledge. While in Tanzania, they trained the doctors on how to effectively use the computers to search for medical information.

The growth of the general program pushed the organizers of the study-abroad program in Tanzania to reimagine its mission. We decided to create a sister program that focused on community and named it Sustainable Service Learning. This allowed for two sections, the Maymester general program and the Sustainable Service Learning, to develop independent logistical plans to maximize the limited time in the country. The latter catered to students who were interested in community service while the Maymester program offered interdisciplinary courses as it was originally designed.

The two programs were rewarding to both the faculty and the students. Students' perspectives changed. They grew to appreciate the opportunities that were available to them at home and which they had taken for granted. They appreciated the hardworking people they met everywhere they went abroad and were impressed by their life stories. They realized that the food served came from the surrounding villages and was brought in by women who were the sole producers and who carried the food on their heads, walking miles to sell it for very little. They wondered whether they could endure that lifestyle and yet have a smile on their face. To share these nuances and their realizations there was a debriefing. Each student was asked to recall one standout thing or incident from that day. A general discussion followed with time for the students to ask questions. These sessions proved to be extremely valuable, with more learning taking place there than in the formal classroom. Emphasis was not on what was told, but on what we saw and did.

The Sustainable Service Learning program was instrumental to my creating a foundation: Sustainable Service Learning, One Kid at a Time. The name was suggested by the students who conceived a sustainable project to support orphaned children who are five years or older and who could not be accommodated at an orphanage that cared for children under that age. In 2018, the foundation embarked on a project to build a permanent home for these children.

The initial buildings were completed in 2022 and the first group of orphans moved into the facility on December 22, 2022. This facility can host forty children and the foundation will provide funding for their upkeep and education until they finish college.

These outcomes of UGA's involvement in study abroad in Africa are fulfilling, and the Tanzania program has remained the most prized activity in my teaching career at UGA and my continuation after I retired in 2018. It is refreshing to meet an alum who often reminds me of the year they participated. I was surprised by one student in 2019 at Emory Hospital. I had been admitted for knee surgery. As I was waiting to be wheeled to the operating room, I saw a young man dressed in green scrubs. As he entered my room, he called out, "Mama Moshi!" Students picked up this moniker in Tanzania from the way local people addressed me, a term of endearment and respect for a woman who is an elder in the community. Hearing that, I guessed this must be someone who knew me. He proceeded to ask if I remembered him, and I drew a blank. He reminded me of the year he was a student in one of the Maymester groups and told me he was now a surgical intern. The icing on the cake was when he told me he was assisting my doctor in the operating room during my knee surgery. I was so happy to see him and to know that he was going to be on my medical team. That was what teaching was all about. The following day, around six in the morning, he showed up in my room to check on me. He wanted to see how I was doing before he left for his other duties. I told him he had made my day.

We have had many other successful students, doctors, and professors, some working in East Africa with Bill Gates' software and health programs as well with Google. When asked, they all said the opportunity to study abroad was the main factor in their success and sparked a keen interest in service. I am grateful to them because they allowed me to contribute, in a small part, to their bright futures.

— 18 —

SHARED EXPERIENCE WITH MY FAMILY

Despite the hard work and successes achieved over the years, I tried not to forget my roots. I reminded myself at every stage that the first source of my success is my family, starting with my parents who made a lot of sacrifices for me to become educated and for valuing girls' education equal to boys. Throughout my journey when I was pursuing education in England and America, I yearned for time with my family, but because of insufficient funds, I was unable to see them for long periods. It took me three years from when I arrived in America to be able to return to Tanzania for a visit. The graduate assistantship funding was just enough for the bare necessities to last the entire calendar year since we did not get paid during the three months of summer recess. It was always a blessing when I got help from unlikely sources like the California Women Educators Association. They gave me $1,000 in scholarship money one summer. This scholarship was a fluke because I did not know anything about them. I had a friend who invited me to brunch at her parents' house one Sunday afternoon. There, I met and talked with other family friends, specifically one guest named Elizabeth, a retired teacher from the Cal State Los Angeles School of Nursing. She asked me how I got to UCLA from Tanzania. Since I am a detailed person, I spent a long time tracing my steps from Tanzania to England, and finally to Los Angeles. She noted that I was resilient and then asked how I financed my studies. I told her that I was a graduate teaching assistant, which provided me tuition relief and a monthly paycheck during the school year but not during the summer. "How do you survive in the summer?" she asked. I explained to her that I am always on a tight budget to make sure I get through the summer and that sometimes I get gifts from friends. Eventually, we ended our conversation.

While I thought my conversation with Elizabeth had ended at the Sunday brunch, I was surprised to get a phone call from her two weeks later. She reintroduced herself and apologized for getting my phone number from my friend's family. She called to invite me to a luncheon that she organized. She also asked me if I would speak at the luncheon about my childhood in Tanzania. Because this would be a short talk, I told her I would just give brief remarks and let the guests ask me specific questions. She agreed. Little did I know, she had nominated me for their summer fellowship for graduate students. The nominating committee had selected me, but she kept it a secret until the luncheon. After my talk and the Q&A, the president announced the scholarship and handed me a check for a thousand dollars. I was stunned but overjoyed at the same time. This was unexpected, more so because I did not know anyone else in this group. Elizabeth and I became friends, and we wrote to each other for many years until she moved to Las Vegas, and we lost contact.

Because I was able to save some of the scholarship funds, I started to put aside money for a trip back to Tanzania to visit my family. The trip became possible during summer 1984. To go home after three years felt refreshing. While in Tanzania, I spent a lot of time talking about my experiences in America, realizing as I talked that I had been away for so long that I needed to readjust and adapt to local conditions. It was strange to feel culture shock in my home country. Everything looked and seemed different: The roads appeared narrower, for example, and doing what I used to do easily had become a hardship. I noticed I was now impatient and could not deal well with how slowly everything moved. I felt at a loss without a car and had to depend on local transportation or rides from strangers. It was even more difficult when it rained, which was something that never bothered me before. Tanzania has red soil, like in Georgia, and when it rained, it became very slippery. If the soil got on your clothes, you would need to soak them for a day or two to get the stains out, and then you had to hand-wash everything. I did not plan to bring back any clothes I wore at home to spare myself the tough laundry task. Nevertheless, the excitement of being home surpassed the temporary hardships.

The homecoming made me realize that I had competing experiences that I needed to evaluate to understand how far I had come in my life's journey. Contributing to this was the fact that my life journey covered three continents, where I was born and grew up (Africa), and the two continents that fostered my continuing education (Europe and America). I settled in America, the place I now call home, but in truth, it is my second home. I realize, as I write this book, that my experiences were shared by my family in their minds, particularly my parents. They were a part of my journey although we lived on different continents.

One of the dreams I had was to bring my family to America to share my experience. I did not know how or when, but my parents were a priority, starting with my mother. I missed growing up with my mother because I only spent eleven years with her at home. The rest of my adolescent years were spent at educational institutions across the country. My father was my grandmother's only son, and for some reason, it seemed like he was the only child, although he had a sister. Because my aunt was married and had her own home, she came to our house to visit and to help when my mother was sick or had a newborn. But every time I saw her, she always seemed like a guest. I visited her a lot at her home, but that feeling didn't change. I still felt like she belonged to someone else's family. My grandmother did not talk much about her, and I regret I did not ask her about her life before she died.

One of the main reasons I wanted my mother to visit was to show her what I had become and to express my deep gratitude for the many sacrifices she made so that I could accomplish so much. Both my mother and all my aunts on both sides of the family did not have the same opportunities as my father or their brothers, to go to school and pursue a career. Both my mother and aunt on my father's side married young and worked on the farm all their lives. My mother had eleven children and my aunt had seven. They both only had two girls and the rest were boys. They feared for the plight of their girls but at the same time believed in the culture that relegated to them certain roles and opportunities. Girls grew up and got married. On the other hand, my father was different from my mother and his sister. He believed that all his children were equal and was very protective of my sister and me. At age eleven, I was the first to leave home and go to boarding school. My sister Matilda also was sent to boarding school at age eleven. My two older brothers, Wenceslaus and Ladi did not go to boarding school until after middle school, well into their teen years. Because my aunt's husband was a builder and did not make much money, my father helped his sister with school fees and uniforms for her children. Her eldest daughter went to the same boarding school as me. My father believed that girls were vulnerable and needed protection. The best defensive weapon, he told us, was education. He was sympathetic to his mother, sister, and wife who had no options.

My grandmother had a very different view than her son: Men are providers and must have all the tools needed, which included a good education. She always asked me how much longer I planned to stay in school. This was when I was in my second year at prep school, the two years required before one could apply to a university. She asked, "Are you planning to be like Nyerere?" Nyerere was, at that time, the president of Tanzania. I do not remember what my answer was, but knowing how witty I was, I must have said something that she did not want to hear. She worried that I was pricing myself out of the marriage market.

With this background, I wanted my mother to see that I had become someone, and that most of all, I was celebrating her with my independence. Because of her support, I had a love for hard work that helped me become a professional. She did not believe that I was really teaching white students. I wanted her to see the University of Georgia and most of all, experience life in America so that when she thought of me, she would have a reference point in her mind. I wanted to tell her that I did what I could for myself and for her. If she had a chance, she could have done what I had done and, who knows, even more.

I wanted to bring her to visit in 1994, but then I took a sabbatical to teach at UPenn. My new plan was to go home for Christmas in 1995 and then return with her to stay with me for five months and then take her back in May when school was out for the summer. They say, "Man proposes, God disposes." That is what happened. On November 4, 1995, my mother died suddenly. She had been ill for some time but, based on my siblings' account of her death, the sudden change in her health was unexpected. She was staying with my sister to have easy access to good doctors in the capital city of Dar es Salaam. That morning, she woke up, took a shower, and packed all her belongings like someone going on a trip. She ate breakfast and, at the table, joked and played tricks with her grandson who was about ten years old. After breakfast she retired to her room and lay flat on her bed. My sister went to check on her and found her gasping for air. She and her husband rushed her to the hospital, but she died on the way.

It is devastating to lose a parent, but it is made worse when you are tens of thousands of miles away. I was able to get home for the funeral, but that round trip was hard. I wept all the way there and back, regretting the fact that I had missed the opportunity to spend time with her. I have never been able to reconcile the fact that she was unable to visit me in the United States.

Afraid something might happen to my father too, I hastened my plans to have him visit, and in 1996, I made the necessary arrangements. I wanted him to do it for himself and mother. I went home for Christmas with a ticket in hand and a visa application for him. We talked, and he agreed to come to America with me for a short visit. I went ahead with the application for his visa, and it was approved within two days. I was not sure how my father was planning for his big trip. He treated each day as another normal day and did not talk about the upcoming adventure. I was only assured that he was thinking about it when he summoned his sons to come home from the city for Christmas.

After lunch on Christmas Day, he made a formal announcement of his plans to travel to America for a few days. He proceeded to give out assignments for each of his sons, all of which dealt with managing the homestead and the farm. There was a disagreement about the assignments, and after listening intently, my father left the room without saying anything. We did not know what was on his mind until two days later, three days before we were supposed to fly out. This was December 27, a special day in the village. It was the annual feast of Saint John the Apostle and Evangelist. This is a special day for the village church because Saint John is the patron saint, revered and much celebrated each year on this date. For reference, this celebration is like the Irish St. Patrick's Day. Because my father's name was also John, our house made it a special day too since we did not celebrate birthdays, but the day Catholics designated for the commemoration of the saint whose namesake we shared.

On this day, families made a lot of food and local beer for adults. The young people got their annual Pepsi, making them feel special. Family friends were invited to the house, but neighbors and extended family members did not require an invitation; they just showed up. They ate, drank, and occasionally got drunk. My father would ask one of my brothers to escort those who appeared too drunk to get home safely.

Those were the days. But in 1996, Saint John's Day was not the same because one important person was not there, our mother who had died in November 1995. Some of my siblings could not be home at this time either. It seemed that for some of them, the need to be home at this time was gone. The 1996 gathering was very small, with just a few of my siblings and their immediate families. My father did not plan to entertain any visitors that year either. The women in the family teamed up to make the lunch. All those who were present had some part to play to make the day special for my father.

After we finished eating and cleared the table, we sat around as usual, drinking and laughing. Dad was unusually quiet that day. I thought he was thinking about the upcoming trip. Then he faced me and said, "My daughter, I want to thank you for arranging a trip for me to go to America. But I am sorry, I will not be able to go this time."

I was stunned. In shock, I said, "Why, Father? I already have your ticket and your visa. We only have three days left. I will lose the money for the ticket and the visa."

"I know," he said, "God will give you more. I do not think it is wise for me to go now. I assessed the situation here at home, and this is not the right time for me to go. I am sorry."

I have known my father all my life, and one thing I was sure of was that there was no way to persuade him once he made up his mind about something. His word was always final. I felt like crying, but I just sat there frozen. No one said a word after that. My father got up and left the room. Our family had two houses: one where we were born and a new one that my father had built. He used a part of the second house as a shop and then changed it into a residence. He spent most of his time in that house. He left to go there, and we did not see him until the next day. I felt that the perfect day had just been ruined. One by one, we stood up and left the room. We did not know what to say to one another and we did not know who to blame for my father's decision.

Early the next day, I contacted Janet in Dar es Salaam. She owned a travel agency and was instrumental in obtaining my father's visa. She knew the process and paid for his paperwork through her agency to make the process go more smoothly. She thought I was calling to have her say goodbye to her uncle and wish him a good stay in America. But she was shocked when I told her that he canceled the trip, and I needed her help to cancel his ticket. The cancellation went smoothly, and I was given a one-year credit to use the ticket. I thought I would use it to come back to visit my father before the end of 1997.

My father did not mention the trip again. I knew better than to engage him in a conversation about it. He was a very reserved person when it came to his feelings, and I understood why he had to make that decision. He was always in control of his plans and made sure he accomplished them his way. He felt this was not the time to make any assumptions about the support he had, and that was why he claimed that the time was not right for him to leave home. I flew out on New Year's Eve. That morning, I made breakfast for everyone, cleaned up the house, and started packing. I had organized a ride for the airport and shortly after lunch, I left for my return flight to the United States without my dad.

Flying back was not easy. There were a couple of times that I wept quietly. I had mixed emotions, remembering my late mother and trying to get over the fact that my plan to have my father visit failed just like my mother's had. I had a strange feeling that this was not over, but I did not know how much time I had to make it work. I was worried about my father, that he would be distraught, slip into a depression, and then give up the will to live. All his life he had never lived alone or cooked for himself. This type of sudden loneliness would change any human being. He held dear, in his heart, the village where he had lived all his life and where he taught, farmed, and raised livestock, activities that kept him remarkably busy all year-round. He knew everyone in the community and could socialize, visit, and talk politics. But the village was not all he wanted, he longed for his family to be with him, and he knew his wife was the missing link.

I arrived safely in America, my second home, on New Year's Day. As usual, I got back to my busy routine. Preparing for the new semester was a good distraction, but I could not forget completely how 1996 ended. I could not shake off the feeling that I had lost a chance to have my father visit me while he could still travel. But luck was still with me. An opportunity for me to return to Tanzania came in March 1997.

I was invited to lunch by Dr. Kathleen Harris and Dr. Sandra Whitney, both of whom worked with the Foundation Fellows program. They wanted me to assist them in preparing the fellows for the trip that was planned to occur in the summer of 1998. We discussed what my role would be, mainly in-country logistics. I expressed the need for me to do a pre-planning trip to Tanzania. Although this assignment was daunting, since it was my first time to participate in a student study-abroad trip, I felt that I was up to the challenge. After the meeting, I contacted Ambassador Mongella to give her a heads-up about the plans and my intention to visit Tanzania in March 1997 for a pre-trip planning and logistics.

The Foundation Fellows office made all my travel plans and I headed to Tanzania as scheduled during the spring break. My intention was to spend time in Dar es Salaam checking sites and meeting with officials who could help me make the program successful. Specifically, I needed to meet with Ambassador Mongella, for a planning session. At the time, she was a member of Parliament. Because of her stature, I was confident that my logistic plans would go smoothly since she was influential at different levels of the bureaucracies. After discussing the plans and the timeframe, she assured me of her full support and asked me to leave the logistical groundwork to her. That was a big load lifted off my shoulders. I gave her a list of what the students were interested in and left Dar es Salaam for my home village to visit with my father for a couple of days.

I wanted to visit my father because when I left after the December 1996 visit, my heart was heavy. I wondered if my attempt to have my father to return with me to the United States had failed just like my previous attempt to bring my mother to visit in 1995. I also felt guilty that, as a family, we all left right after my mother's funeral, and my father was alone in the house for a whole year. No one had visited him until I returned in 1996 with a plan to bring him to the United States. He had looked feeble, which worried me a lot. I was eager to see him on this short visit to assess his wellbeing.

I arrived in the village late in the evening. My father was waiting for me because he knew ahead of time that I was coming. I had stopped in town and bought some groceries. Once I settled in, I began to make dinner. Luckily, one of my brothers was also visiting home, so we had a nice, small family dinner. Afterward, we lounged around talking about world politics. Suddenly, my father said, "My daughter, I am ready to go to America now. I will go back with you."

I was in shock. I did not want to show it and pretended that what he said was fine. Deep down, I did not know how I could make it happen in two days. I said, "I will see how we can get a ticket. If I am not able to do that, I will arrange for you to travel and I will meet you in America."

He looked at me and said, "I want to travel with you. I am ready."

My father thought a trip to America was like a trip to one of the cities in Tanzania, where you just pick up, go to the station, buy a ticket, get on the bus, and off you go, and in a few hours, you are at your destination. Again, I knew better than to argue with him. "I will work on it," I responded. He kept quiet for half a minute and then raised his head, looked at me, and said, "Thank you, my daughter."

I went to bed feeling restless. I could not sleep because I knew time was of the essence. I could not extend my stay because I had to be in class in a couple of days. Early in the morning, I contacted my cousin, Janet, in Dar es Salaam and told her what had transpired. We already had the visa, so that was easy, but I did not have a ticket for my father. She promised to work on it and let me know as soon as possible. Janet was a genius and had a sweet tongue that could lure a snake from a cave. She called KLM, the Royal Dutch Airlines, to retrieve the ticket that we had placed on hold for a year. She was successful, but it did not look likely that we could be on the same flight all the way to the US. Janet pleaded and shared information about my father's age and the need for him to travel with someone he knew. They gave her time to find another flight where my father could be with me. Because Janet could track the seats on the KLM flights, she checked on it every half-hour. Suddenly a seat opened on the same flight I was on, and she reserved it. She called KLM and told them she had a seat she could give to my father. This is the power of those who know how the system works. As a travel agent, Janet could negotiate with the airline. Around three in the afternoon, Janet sent a message that all was well, and my father and I could travel together. Quickly I set out to find my father, who was at one of his gardens. I told him the news and encouraged him to begin preparing because this was going to be a month-long trip away from home. I thought he would show some emotion about the news, but he looked at me and casually said, "Fine." After the news, my father did not mention the trip or ask questions. At first, I was nervous that he was not preparing at all for our trip. The fateful travel day arrived. As usual, he went through his routine of cutting grass for his goats, feeding his cows, and other small tasks as if he were not going anywhere. I was nervous about him reserving enough time to get washed, packed, and ready by the time we were supposed to leave. To my surprise and relief, he was ready to go when the driver pulled up.

On the way to the airport, my father was extraordinarily quiet. The driver helped us with the luggage and said good-bye. My father walked with me to the check-in counter. I handed over both passports. The agent tried to engage my father in a conversation.

"So, Mr. John Semali, you are leaving us for America, the big nation?" he asked him.

My father just stared at him and said nothing.

"Okay, *Mzee* (a term of endearment for an elder statesman), have a great trip, and enjoy your stay in America."

Still no word from my father. I thanked the agent and moved on to immigration clearance. We went through immigration and the security check very quickly and moved on to the waiting lounge.

"Oh, a lot of people are traveling. All going to America?" my father asked.

"I do not think so, Father. They are traveling to different parts of the world. We are first going to Amsterdam, then we will change planes and go to America," I replied.

"Okay," he said.

Shortly after, the KLM flight arrived from Amsterdam. After another half-hour we were called to board. Because of my father's age, we were among the first ones on. It was dark outside, but you could see the plane on the runway. My father seemed surprised by how big it was. He had never seen a plane on the tarmac, only flying above the banana and coffee trees, leaving a trail of smoke behind it. He was about to see how it looked inside.

The attendant showed us our seats. I told my father to sit by the window for a better view after we took off. I detected worry in his eyes. I am sure the idea of flying in the air like a bird was daunting to him.

Once everyone was aboard, the captain announced our departure. The flight was going to Dar es Salaam, the capital city of Tanzania, to drop off cargo and passengers from Amsterdam, and get cleaned, and refueled. Then new passengers and cargo from Dar es Salaam would be loaded before it was off to Amsterdam and then to the US. As the plane started to taxi, the noise bothered my father. I could see he was scared. I told him it was going to be fine because I had done it many times. I told him that the noise was coming from the engines, and that it was normal and safe. He trusted me. I told him once we were in the air he would not hear or feel anything. He could get up and walk around, just like he was in a house. He trusted me enough to believe me. I started a discussion on politics to distract him and this worked. Once we were above the clouds, the noise did not bother him anymore. It was a moonlit night and when the captain told the passengers to look through the window to see the snow on Mount Kilimanjaro, I opened the window shutters and pointed it out to him. He was utterly surprised. He started talking about it, and I could see that he was much more relaxed.

We arrived in Dar es Salaam after about forty-five minutes. Those continuing to Amsterdam were asked to remain on the plane and all the others had to disembark. The cleaning and stocking began. My father was amazed. Shortly after the new crew arrived, passengers followed. Around midnight, the plane took off again. This time, my father was not as scared by the engines' roaring. Once settled in the air, the host came around to take drink orders. My father ordered a beer, and I ordered a glass of wine. Then came the meals. I chose beef and rice for him. Later, I escorted him to the bathroom because I knew he would be scared walking to it. Once I was sure he was comfortable, I waited outside. He came out and we walked back to our seats, I held one of his hands while he supported himself on the seats as he walked back. This was truly an adventure. I was beginning to wonder how he would hold up for the entire eighteen hours of the trip. I ordered an extra blanket for him and helped him recline the airplane seat to get some sleep. He slept throughout the night and was awakened by the call for breakfast. I was glad to get off the plane in Amsterdam to allow him to stretch and see a part of the world he had never even heard of. As we were walking to the gate, a young man approached me and asked how to get to Gate 15. I told him we were heading there too, so he could walk with us. He seemed relieved. I asked him where he was going and learned that he also was heading to Atlanta.

"I am going to Georgia Tech University, to start my master's degree in computer science," he added.

"Where are you from?" I asked.

"Tanzania." He answered.

"Oh!" I exclaimed. "This is my father, John Semali. We are from Tanzania and going to Atlanta too. I am a professor at the University of Georgia," I added and switched to Swahili.

"Your school is called Georgia Tech, a way to distinguish it from the many schools in Georgia," I added.

This was a surprise to him coming from a country where there was only one university. He exchanged greetings in Swahili with my father. This was a blessing because now I had some help for a little while. I asked him to keep an eye on my father in the bathroom and help him if he needed anything.

After the bathroom visit to freshen up, we headed to Gate 15. We had an hour before the departure. Anakleti, the young man, spent most of the time talking to my father. I am not sure what they were talking about, but I was glad to see them so engaged with each other. They were two people from different generations who were experiencing their first trip by air to a foreign country. My father was no longer scared. Now, there was someone else who spoke his language, shared his culture, and was having the same experience of being away.

Eventually, it was time to start boarding for the last leg of the trip. I took advantage of my father's age to board early once more and asked if the young man could come with us. The attendant thought we were one family and let us in. After settling down, I caught my father's eye. He was mesmerized by the countless number of planes he could see outside the window.

"Where are all these planes going?" he asked.

"All over the world," I replied.

The usual plane services began as soon as we reached the desired height. After lunch, my father started dozing off. I found an extra blanket and pillows for him to make sure he was comfortable. Once settled, I watched movies. Throughout the entire trip, my father got up once to go to the bathroom. By this time, he was familiar with the system and did not depend on my assistance. When we arrived in Atlanta, he looked very tired. He also looked overwhelmed by the number of people going through customs and immigration checks. Luckily, the process went smoothly, and we were out in about half an hour. My friend Frances Rees came to pick us up at the airport for our seventy-five-minute ride to Athens. I met Frances when she was a graduate student in Adult Education and asked me to be on her dissertation committee. Later, we volunteered together in the Jefferson County Correctional Facility in a program on culture and personal responsibility. She came to my rescue when my mother died. Because I got the news on the weekend, I could not get a ticket for my trip in time for the funeral. She volunteered to help, and through her friends, she got me a ticket and paid for it with her own credit card. I paid her when I got back, but she returned the check telling me that she and her friends wanted to cover the cost. I remember calling her and sobbing because I could not believe that such generosity existed, and from people I did not even know. When I told her I was coming back with my father, she immediately volunteered to meet us at the airport and drive us home. She lived in Commerce at that time and owned a large farm with horses and wanted my father to visit. I made the arrangements after my father got settled. He was impressed, more so because a woman owned a farm.

During the ride, I thought my father would be dozing off, but he had his eyes fixated on the route to Athens. He was quite surprised by this new place his daughter called home. He asked about the cows and sheep he saw on the farms along the way. He was surprised that the owners were individual citizens and not the government. In Tanzania, the government usually owned animal ranches and large farms.

At my house, we were met by my brother Adelin who was already in the US and living with me. Although I had made this trip several times, I was exhausted and could not wait to have a warm shower and go to bed. That was what I did after dinner. We arrived on Saturday, and I had Sunday to get over the jet lag before classes on Monday. That gave me time to get my father settled. On Sunday, I took him for a tour of Athens and the university. He was amazed and impressed by everything he saw. I was glad Adelin was there to answer most of his questions about America.

A week after my father's arrival in Athens, he expressed a wish to visit with his friend's son who lived in Florida, a pastor at one of the churches in St. Petersburg. After contacting his friend's son, we planned a weekend visit, leaving Athens on a Friday and returning on Sunday, in time for me to be back in class on Monday. This was my second trip driving to Florida, the first one was to visit a colleague who left UGA for Florida Atlantic and lived in Boca Raton. Nevertheless, I was a bit concerned about driving for ten hours, but I wanted to fulfill my father's wishes. I had a good car, a Toyota Cressida, which could take me that distance on a full tank full of gas without stopping to refuel.

At six in the morning that Friday, I had the car all packed with our weekend bags, snacks, and water. I had a pillow for my father just in case he wanted to take a nap during the drive. My father was an early riser too, and he was ready to go. Adelin got up to see us off. As we got into the car, my father turned to me and asked, "Who is driving?"

"I am," I proudly replied.

The next question got me thinking about my father's protective nature.

"Your brother is not going?"

"No, he has to go to school, he has classes."

My father looked at me, looked at Adelin, and I could see the concern in his eyes. I told him I had a pillow for him in the back seat and that he could get comfortable there. Quickly, he responded, "No, I am sitting up front."

I moved the pillow and reclined the seat a little bit before he got settled. I got in the driver's seat, and we were ready to go. I felt like a pilot on a plane, in control and confident. My father still looked worried, and I was not sure whether he was worried that I could drive him or the fact that whenever we went somewhere Adelin was always with us in the car, and now it was different. It is a good thing I did not tell him how long it was going to take us to get to St. Petersburg. He never asked until later in the trip.

I had my road map laid out because there was no GPS at that time. We said good-bye to Adelin, and I started to pull out of the driveway to the main road. Discreetly, my father took out his rosary and started praying. I do not think he realized I saw him do that. He had his left hand firmly planted on the dashboard as if holding it from falling off. I tried to assure him it was going to be a good trip. The sun was up, and he was going to see a lot of interesting sights along the way. I also told him he could relax his arm and rest because the seatbelts were strong and would protect him. He realized that I was aware of his fear, and he did what I had told him to do. He wanted to assure me that he was fine and trusted me.

    Once we got out of Athens, I took Highway 441 through Madison, Milledgeville, and another smaller town. At 8 a.m., we stopped at a Waffle House for breakfast. By now he was much calmer. We finished eating and headed for Highway 75, which was going to take us the rest of way. I tried to distract him by asking questions about his childhood and life. This was the best idea I ever had. I learned so much about how he grew up as the youngest son of his father and the only son of his mother who was my grandfather's fourth wife. During his era, men could marry as many wives as they could support. I learned about his relationship with his sister, who did not go to school, while his mother sold one of her cows to send him to a boarding school run by the Holy Ghost fathers. He considered himself lucky because when everyone was doing the entrance exam, he and a friend were sent by the school principal to the farm to collect some supplies. Because they were late for the exam, the two boys were admitted without it. The explanation was that they had completed a task worthier than the examination. He said he was fortunate because both boys came from a primary school that was not known for producing successful students. My father thought that if they had taken the exam, they would have failed and then been sent back home. His father would have been happy about that because he adamantly opposed his son's enrollment in boarding school.

"I do not trust the white man," my grandfather said. "Their plan is to take our boys away from the village and turn them into little white boys who do not respect farming and herding."

That is why he refused to pay his school fees, and my grandmother, who was determined to see her only son become someone, sold her best cow so he could go to this white man's school. Hearing this story was not only fascinating but answered the questions that I had had. What surprised me was that my father was the only son of my grandfather who had an education beyond the fourth grade. The chance for an education and to become a teacher was the main reason behind my father's passion for education for his eleven children, regardless of their gender. The first question he asked us whenever we came home, also from boarding schools like the one he attended, was, "How was school? What are your grades? Do you have a report for me to see?" There was no opportunity to lie or fudge the answers. Bad grades meant less pocket money for the following school term.

His stories kept my father occupied and relaxed for a good two hours. I could tell he was enjoying telling me about his childhood and growing up to be as successful as he had been. I understood why education for all eleven of us was so important to him. I wondered why he was so scared of us failing to succeed. Apart from the coffee and his meager salary from his teaching job, he had nothing else with which to support our education. I also thought his fear was the possibility of all of us not leaving the nest and being at home with no hope for the future.

Highway 75 was busy that day. Every time a car drove past us, he would say, "These drivers drive very badly. Why are they going so fast?"

I responded calmly, "This is a highway where drivers can go fifty-five miles per hour, but some drivers go over the limit. The police will get them," I concluded. The limit had not changed to sixty or seventy-five like today.

My father seemed amused by the thought that there were police on the road to regulate speed.

"Where are they running to?" he asked.

"Different places, Dad," I said, "Some close, some far, like us."

A few minutes later he asked, "So, how far are we going?"

He finally realized that the trip was not as short as he had thought.

"We have a few more hours to go," I replied. "Do you want a bathroom stop?"

He paused for a moment and then replied in the affirmative. After about five minutes, I saw an exit for a rest stop, and I took it. This was a good place for a break and to stretch.

After the short break, we embarked on the remaining miles to our destination. We arrived at four in the afternoon, the predicted time. My father was delighted to see his friend's son but most of all, to come to the end of the trip. The first thing he said to him was that if he had known how far we were coming, he would not have asked me to do it. In addition to him being tired, he was sorry I had to drive for so long. Our host came to his aid and said, "Oh, do not mind her, she can take it, she is still young."

After getting our luggage to our rooms, we settled down for a cup of tea. Because the pastor was Catholic, he lived in this house by himself. I realized that my primary culture was going to be tested. In my acquired culture—that of the US—I was a guest, but in my ancestral and traditional culture, I was the only woman in the house. As a guest, I would have taken it easy and been waited on. Tired or not, I found myself assuming the woman's role of providing service where it was needed. I cleared the teacups and took them to the kitchen. My father and the pastor continued to chat, catching up on news from home. While I was washing dishes, the pastor came to the kitchen and said, "I am so glad you could bring Mr. John to see me. I am also glad to see you. It has been a while. Look at you, a fine, elegant, and smart professor teaching in the United States."

Politely, I responded, "Thank you, I am glad to see you too. How have you been?"

"Thank God, very well, in good health and working hard," he replied. "This is a big parish and a lot of work. Anyway, we will talk more about that later."

Before turning to go back to the living room where my father was, he said, "Do you know what I would really want to eat today? Banana soup and flatbread.

He proceeded to show me all the ingredients that he had bought ahead of time for the dinner in the refrigerator. This included meat, flour, green bananas, tomatoes, onions, rice, and cabbage. I did not have to say anything because I got the message.

"You are now in charge of the kitchen, and good luck. See you at the dinner table."

I started preparing the dinner. I planned to finish before seven so I could retire early. Little did I know that he was planning a party. About five minutes later, he came back to the kitchen and said, "I forgot to tell you, there are three nuns from the parish who will be joining us for dinner. They are from Tanzania and are going to school here."

I could see that my plans to retire early did not align with the events for that evening. I had to plan a dinner for six. Making the banana soup and rice wasn't too much work. The difficult task was making the chapati flatbread. This was labor intensive and took a great deal of time without help. To make chapati flatbread, one had to start with making dough like you did for pizza. By my estimation, dinner wouldn't be ready before seven-thirty.

I was glad I liked to cook, and no amount of cooking bothered me once I had a good plan. I decided to start making the dough first because it required an hour of sitting to get soft before rolling it out. Then I proceeded to make the soup. While it was cooking, I made the sauce and the cabbage. About an hour later, the pastor came back to the kitchen to see how I was doing. He was impressed by my speed and said, "This reminds me of watching you at your father's house when we came for Sunday lunch. You and your sister were such gracious hosts, well-mannered and educated by those nuns from England."

I smiled in acknowledgment. I was not in any mood for small talk, so I kept chopping the cabbage and sauteing the meat for the sauce. Before he left the kitchen, he declared, "Oh, I do not want to forget. My niece just called, and she will be coming with a friend too."

"Okay," I replied. "Just so you know, it is too late to change the portions, but I think I made enough to feed an army, so no one will go home hungry."

He did not say anything else and left the kitchen. This was great because I could concentrate on what I was doing instead of thinking about what to say in the middle of this intense meal preparation. However, I was glad his niece was coming and bringing a friend. I knew that culture dictates that age comes before beauty. Since they were younger, my responsibilities would soon be ending with the cooking. I was not sure about the nuns, even though they would be younger than me too. Because they were "women of the church," there was no way I could expect them to just roll up their habits (the outfits they wore) and take over cleaning up after dinner. But with the arrival of these two young girls, I was free to relax.

I had planned to finish cooking before the guests arrived, which worked perfectly. At seven-thirty, I was done. I had time to take a shower and freshen up. I joined my father and the pastor in the living room where they were enjoying a beer. There was a bottle of wine on the table. I took the liberty of pouring myself a glass and settled in an empty chair near the window. Before I had two sips, the doorbell rang. The pastor went to open it and the two nuns came in. They greeted my father and introduced themselves. Because my father had been a teacher for so long, he knew everyone in our village and the surrounding villages. He proclaimed that he had taught both of their fathers and went on to praise them as particularly good in mathematics, the subject my father taught. The pastor offered them a drink and they chose the wine. The glasses were already on the table, but I had to get up and serve them. Somehow, I started feeling once more like the "head of the house." Shortly afterward, the two girls arrived. They were familiar with the house, and I was happy to let them take over the service part of the evening. They laid the table while I put the food in the various serving dishes. Once the food was on the table, it looked impressive, like a feast. The pastor said grace. It was a long rendition of prayer items, remembering parents, relatives who were near and far, and everyone present, particularly my father. He spent some time on him, thanking God for his accomplishments in raising us and giving us an education, my father's friendship with his father, his long trip from Tanzania and to St. Petersburg, and a safe trip back to Athens and Tanzania. I am sure everyone was happy when it was over because the food was getting cold.

Dinner took a while to finish; we were not done until around nine-thirty. I did not mind because I was convinced that my job was done and as soon as I could steal away, I was going to bed regardless of whether the party was over. That is exactly what I did. I did not tell anyone. I just disappeared from the room. It was around one o'clock in the morning when I woke up and heard conversations still going on in the living room downstairs. My father and the pastor were still talking. They must have been cracking jokes or something because the laughter was loud and intense. I went back to sleep and did not wake until around eight-thirty that morning. The pastor and my father had left the house to go to Mass. They did not wake me, and I did not mind at all. I did not know how the pastor wanted to handle breakfast. They came back around a half-hour later. I was drinking a cup of tea at the kitchen counter.

"How was your night?" the pastor asked.

"Very well. How was yours? I am surprised you are up and about after having gone to bed so late last night," I said.

I greeted my father who did not seem amused that I had slept in and missed a chance to go to a Mass that was within walking distance. I tried to ignore all that, and asked, "What are the breakfast plans?"

"I will tell you in a minute," the pastor replied.

He went upstairs to change his clothes. When he came back, he announced we were going to a breakfast place in the neighborhood. I was relieved. No cooking, no dishwashing. It was within walking distance, and we did not need the car. Everyone at the restaurant knew the pastor. I was introduced to at least a dozen people. The breakfast was extremely good. It was a little foreign to my father, but he enjoyed every bite of it. We had so much to talk about, mostly about Tanzania politics, which my father loved to do. After breakfast, we left the restaurant to walk back to the house. The pastor suggested that we pass the church on the way home so he could give us a tour. We met a few of the staff at the parish who were busy preparing for Sunday service the next day. After the tour, the pastor offered to take us for a ride to see the neighborhood. This took most of the afternoon. Because we ate so much at breakfast, we did not stop for lunch but had a snack and soft drinks. We came back home early because the pastor had to prepare for Sunday. With the leftovers from the dinner party the night before, I did not have to cook much that day. The pastor had bought a leg of lamb that we put in the oven and made some vegetables to add to the leftovers.

After dinner, we all retired early. Sunday was a busy day for the pastor. The church had a service at eight o'clock and another at five o'clock. And we had a long day of driving ahead of us. We were planning to start back around noon. We went to the early morning service after which we had breakfast in the church dining room. We met many of the people who worked with the pastor and more nuns who served other parishes. After breakfast, we headed back to the house to get packed and prepared to leave. A little before noon, we started our trip back to Athens.

The return trip was smooth. My father seemed more relaxed, though perhaps a bit anxious about the many hours of travel. But, when you know where you are going, the distance seems shorter. I thought we would be getting home late at night, but we beat the clock and arrived in Athens around seven-thirty. Because we only made one stop, it took us exactly eight hours. During the trip back, I answered more questions about what my father was seeing. He was amazed by the large farms we passed along the way, the herds of cattle, and the lush vegetation. He made comments about how much land was wasted in Tanzania, where people could farm and keep animals. If America could do it, so could Tanzania. He then concluded, "I wish I was younger; people would be surprised." He got some brilliant ideas about what he could do with the land he owned and where he only farmed cereal once a year. He could augment his banana and coffee production with diversified farming and large-scale cattle herding. I wondered if he was remembering his father's words about the kind of education the children were getting. My grandfather thought that the education system divorced children from the fundamentals of a thriving community and, instead, created a community of office workers and city dwellers who depended on other people to feed them. My father was wondering what could have been for him, with eleven children, if we all worked on the land rather than moving across the world while the land, he had acquired was lying fallow with no one to farm it. I felt his pain, but I was glad that he had given me the gift of an incredible education that allowed me to choose my life and not have to live the life of my grandmother, mother, and aunts. By the time we returned to Athens, I was as glad as he was that we had completed our weekend getaway safely. I was grateful that my goal to bring my father to America was a success. My father had the opportunity to see the great country I often talked about, and by the time of his visit, had been my home for more than a decade.

The remainder of my father's stay in the United States went very quickly and when school ended for summer break in May, I went back to Tanzania with him. The visit left an indelible impression on him and left me with gratitude and a feeling of accomplishment that at least one of my parents had an idea of how far I had come.

When we arrived in the village, many of his friends came to see him, curious about how he looked and eager to ask him about his experience. They had endless questions.

"Mr. John, you look ten years younger. You should have just stayed there. Why did you come back?" they asked.

As usual, my father had the most interesting answers. "Your home is your home, even if it is a foxhole," he replied.

He talked endlessly to them about America. In fact, it was the topic of conversation all the time to anyone who would listen. I was not with him during the first few weeks of his return to the village because I had students to teach and manage in my study-abroad program. I visited with him for a couple of weeks after the program concluded and before I returned to the United States. The villagers told me that his visit to America had changed his philosophy of life. They also commented on how proud he was of me and my accomplishments. He was happy that I had a stable, independent, and secure life. Finally, he was confident that I could stand on my own two feet and that he did not have to protect me or worry about me. For him, this was unusual, and this woman happened to be his daughter.

As happy as I was about fulfilling my dream of bringing my father to America for a visit, I still regretted that my mother did not have the same opportunity. I wanted to tell her that I became who I am because of her, and my accomplishments are her accomplishments. She was denied the opportunity to be all she could have been, but all was not lost because I achieved what I have done for her. I remember her crying for a week because I left home to go to boarding school. I always wondered whether she cried because she was going to miss me, the first child to leave home, or if she was scared for me since I was only eleven years old. It was for both.

After my father's first trip in 1997, he was open for a second visit in 2003. This time, he traveled with my brother Cassian who had gone home for a visit over Christmas. He did not intend to stay for more than a month and was convinced he could make his way back alone. But this wish was not granted because before the month ended, he had a stroke.

When the stroke happened, I was at a conference in Montana. I knew he was not feeling well, but I was not aware he was so seriously ill. When I got the news, I quickly arranged a flight back to Athens and found him in intensive care at Athens Regional Hospital (now Piedmont Athens Regional). The thought that my father might not live through this ordeal was unbearable. Miraculously, he recovered enough to be transferred from the ICU to the cardiac ward at the hospital. He spent a total of three weeks in the hospital. To walk again, he endured what seemed like endless physical therapy. He also temporarily lost his memory, often thinking he was home in Tanzania. He asked about people he knew, some already dead, and why they had not come to visit him.

He also asked about his livestock, wondering if they were being fed properly and cared for as he would have done.

The most painful part for me though was watching my father undergo these transformations while living in the hospital. It also was difficult coming home in the mornings to change before the office, then going to class to teach, and then returning for another night's stay at the hospital for three weeks. The day he was discharged was the happiest day of my life. His doctor had recommended taking him to a rehab facility, but I declined, insisting that I would be able to manage him at home. The doctors were not sure about this decision because he still needed help washing and dressing. But with my two brothers, Cassian, and Adelin, in Athens, I was convinced we could manage. I was sure my father would not do well in a rehabilitation center but would recover faster in a familiar place—my home in Athens. I was right, because in three weeks he made tremendous progress. He was able to wash and dress and take short walks around the neighborhood. When we went for his doctor's checkup, everyone was surprised to see him walking by himself without a cane or walker. The stroke did not affect his limbs or speech. He ended up staying in Athens until October because he underwent cataract surgery on both eyes. The surgery's success gave him a new lease on life because he resumed his love of reading. This was his last experience in America because he never had an opportunity to return. He died in early March 2007, three years after his last visit.

Although I know that deaths are inevitable and I have gone through the experience of losing someone before, each one comes as a shock. When I received the news about my mother's death, I felt helpless and confused. Learning about the death of a family member is debilitating when they are eight thousand miles away and getting there on time for the funeral seems impossible. The journey home to my mother's funeral was the longest flight I have ever experienced. I cried the whole time. The flight attendant was concerned about me and was not sure what was so upsetting that I refused to eat or drink anything on the plane. She dared to come sit by me and ask me what was wrong. I told her my mother died and I am going to her funeral eight-thousand miles away. Throughout the trip, I constantly felt guilty, blaming myself for not being there.

My father's death was more debilitating because I had just come back from Tanzania two weeks earlier. My brother Venance had died after a long illness, and I had traveled to Tanzania to attend his funeral. Although my father had been ill off and on, I did not expect him to suddenly die. I attributed his sudden death to the unbearable grief of Venance's death. It is believed that children should bury their parents, not vice versa. The grief must have taken away my father's will to live after we all left him alone at home and returned to our respective jobs. I was deeply hurt by his death because on the day I was traveling back to the United States after my Venance's funeral, my father was nowhere to be found. I left without saying good-bye. I tried to call him several times from the US but was unsuccessful. With this memory weighing heavily on my mind, the trip back to Tanzania for his funeral was extremely hard. I had so many "what if" thoughts, which were hard to process. I cherished words of consolation I received later from my doctor who knew him, who said that I should celebrate his eighty-seven years as a life well-lived. I was glad he had visited me in the US twice because I am grateful for the opportunity to remember him in my home. Despite some trying moments,

I was so blessed to spend that extended period with him. He met my friends and many of my new family members who have joined me on my life's journey.

2002 UGA—Tanzania Maymester program students visiting the high school I attended (1968-69).

2003 UGA inaugural Service-Learning program in Tanzania.

At the Geneva Women's Conference in 1990.

*2009 inaugural USA K-12 teachers
Fulbright program in Tanzania.*

Named University Professor 2007. At the award ceremony with UGA President Michael Adams (right) and Provost Arnette Mace.

Award ceremony with (left) Dean Garnett Stokes, Arts and Sciences; (right) Betty Jean Craige, Humanities Center Director; nephew Chris, 5; and other family members.

Florence Millington Heath Snyder, Director of African Studies Lioba Moshi and Darl Everett Snyder

With Dr. Lioba Simon Schumacher (German), the first namesake I met (2012) outside of Tanzania.

UPENDO One Kid at a Time Center buildings for orphaned children (opened December 23, 2022)

Inaugural 23 children at the center (December 23, 2023).

— 19 —

Epilogue

Having settled in America, I have had a chance to evaluate my three different experiences in Tanzania, England, and in America. Each has been an integral part of my life. When I left Tanzania for Europe, I was naive about England. I thought that life there would be simpler because of my familiarity with the culture. I was born in the latter part of the British administration in Tanzania, which afforded me a certain amount of knowledge of the British system. I thought my proficiency in English and familiarity with British culture and etiquette would be an advantage in adapting. However, I quickly realized upon arriving that some of my assumptions were incorrect. The closest contact I had with the British people was with the nuns who taught me for ten consecutive years. I soon realized that the nuns did not live like ordinary people.

I was also naive about racism in England. I did not realize at the time that many of the nuns' microaggressions were rooted in racism, but now looking back, I can clearly see it. The fact that the dark-skinned girls worked in the garden and helped feed and clean the pigs while the light-skinned girls worked with the white nuns indoors washing, ironing, and mending clothes constituted discrimination based on the color of one's skin. There was another incident at York

University when I was singled out on my first day in class. That my professor specifically chose me, alone, to do a linguistic transcription of his lesson no longer seems like an innocent coincidence. I had started school two weeks late and had not yet learned that lesson like the other students. My limited knowledge was tested on the spot, which diminished my confidence and revealed my ignorance in front of all-white British classmates who I had just met that morning.

Another memorable racist incident happened when I was invited to dinner by a host family. Their son asked me if my people lived in treetops. I did not blame the child but the education system that overlooked the fact that the jungle was a source of wealth that contributed to the British economy and that the British had lived in this jungle until its citizens demanded their independence. Often, I was reminded to be grateful for the opportunity to go to school with the British students. I think anyone who has been given the gift of knowledge would be and should be grateful.

I remember one present given to me by one of my students, a coffee mug with the inscription: "To teach is to touch hearts." Therefore, we should be grateful for knowledge and those who offer it. This way of thinking contributed to my survival in England. I knew I was privileged to be there, but at the same time, I had a right to knowledge like anyone else. My goal was to succeed and open the door for others so that they too could have access to knowledge.

I also was fortunate enough to create new families while abroad. I am thinking here of my friend Vera McIntosh who I met while she was visiting Tanzania. It was refreshing to know that I was not alone and abandoned in a foreign country. Vera and her family came from Guyana, and they understood the plight of a foreign student who knew no one and no place to call home. They provided the home and the family I needed as I struggled to achieve my goal. They are still in England and have remained my "forever" family. In 2018 and 2019, I visited with them in Guyana to be introduced to their ancestral home, formally known as British Guyana until it attained independence in 1966.

And what about America? America continues to fascinate me ever since my first visit in 1978 to train the Peace Corps for Tanzania. Countless immigrants have made America their home away from home. One such individual is General Colin Luther Powell, whose success story is exhilarating. Powell, son of Jamaican immigrants, was born in Harlem, raised in the South Bronx, was an undergrad at the City College of New York. This afforded him an opportunity to join the Army and serve the United States in the Vietnam and Iraq wars. He rose to the highest ranks and became the first Black American to hold posts with the Office of the Secretary of Defense, Brigadier General, and eventually became chairman of the Joint Chiefs of Staff. He received the Presidential Medal of Freedom twice, from President George H.W. Bush in 1991 and from President Bill Clinton in 1993. He served as the US Secretary of State under President George W. Bush. I admire his success story, which is chronicled in his best-selling autobiography, *My American Journey*.

I do not think Gen. Powell planned all these accolades, rather, they were a result of his hard work, focus, and staying true to service. These virtues undoubtedly were impressed upon him through his upbringing. In a television interview, Powell noted that: "Black people have come a long way. America has come a long way in race relations, but more must be done to eradicate all pockets of racist thinking and doing." While he acknowledges the struggles and challenges due to his race, Powell has not been preoccupied with how people treated him in the past but has focused on how to lead by example and espouse humanity and respect for all people. By establishing a charity, America's Promise, he has dedicated his service to inspiring young people. He is an example to emulate, an encouragement to any person who wants to make a difference and improve the lives of others. I see him as a source of inspiration for finding a way to help those not as fortunate as we are. That is the source of strength that drove me to establish the One Kid at a Time Foundation to support displaced and orphaned children like those at the UPENDO Center.

I tell myself that these kids did not ask to be born but they were thrust into the circumstances they are in. We cannot blame their parents because they could be victims too. I can only offer help to try to break this cycle. I know it is a drop of water in a huge bucket, but a drop is better than no water at all.

Author with one of the orphans.

I learned throughout this journey that life's challenges and struggles are not meant to serve as a source of discouragement. Instead, I see them as opportunities to find creative ways to solve complex problems. There is a Scripture that I often quote whenever I have a hardship to deal with: "In a little while you will see me no more" (John 16:16). Sometimes I said this to a person, and other times, to myself. When said to another, it sounds like a threat, or at the very least, a concern. However, I have never meant it that way. This Scripture's message is about the problem not the person. I was making a promise to myself not to dwell on my struggles or hardships for long. I convinced myself that a problem or anything I was struggling with would remain with me for as long as I continued to entertain it. But if I focus on finding a sustainable solution, I can overcome it.

Sunset has a calming effect on me. I tell myself that I have eight to ten hours of sleep to not think about anything, and I promise myself not to carry yesterday's challenges or disappointments to the new day. It's not always realistic, but it makes it easier when I have a busy day ahead to avoid focusing on myself. Any success I achieve gives me renewed energy to do more or push harder. This is how I relate to what General Powell said. He did not say he was done once he crossed the first hurdle, but rather, he waited for and approached the next hurdle with new energy and the vigor to knock it down.

As a teacher and administrator, I certainly experienced many struggles that could have deterred me from moving forward. True to my philosophy, I discovered that such detours could often become useful and inspiring. Many of the obstacles I encountered were because I was new to America, an immigrant like Maud Ariel or Luther Theophilus Powell, General Powell's parents. The lack of adequate knowledge and experience of how the American system worked made it difficult to navigate the systems that others took for granted. How do you open a bank account in a foreign country? How do you apply for an identification card or a driver's license? Where do you find insurance for your house or car, let alone your health? All these were new experiences. How do you develop support networks since you have no local family to rely on, people from whom to seek advice? How do you establish a name for yourself as a scholar when you do not know the nuances of the academic network? I had no model in America to rely on. It took time to figure out the basics.

There was a time when I was new to America and realized I needed to develop my support system. One night, I accidentally locked myself out of the house I had bought just a few months before. I did not realize I was locked out until I returned home from a party at ten o'clock that night. I frantically searched my purse and could not find my keys. I could have slept in the car until morning and then sought help, but I decided to drive to my friends' house with the hope of crashing on their sofa until dawn. Luckily, they were still up when I got there. Instead of giving me the sofa, they called a locksmith. I did not know there were people who did that line of work and what's more, operated around the clock. They found one who agreed to meet me in about twenty minutes. I drove back home, and we arrived at the same time. In less than a minute, he had my door open and charged me a hundred dollars. Luckily, I had cash on hand because he did not take checks. I learned my lesson that one can find help even in the middle of the night if one asks.

Other challenges that failed to defeat me were credit cards and renting. It seemed strange that to obtain a credit card one must first build a credit rating. The credit card companies wanted you to have a certain income to qualify unless you had a co-sponsor. My lack of familiarity and experience made it difficult for me to understand the explanations. Eventually, it became pointless to ask for assistance without seeming stupid. In the 1980s and '90s, the internet was not as sophisticated as it is now. There was no Google or Wikipedia to ask questions and receive instant answers. It became obvious to me that creating a network of friends and community was key to survival. From California to Georgia, regardless of where I lived, I had to create alliances, and once I felt comfortable in that social community, I could ask about the necessary information. These challenges emboldened me and taught me how to be a self-reliant American. Considering the strength of my alliances and community of friends, I broadened my life's journey with different kinds of family members, without which I would not have survived, let alone thrived, in the United States.

My time in America has allowed me to open my lens wider to see beyond the horizon and appreciate what is possible when one puts their mind to it. As I look back, I think that despite having doubts about what was possible, I accomplished more than I ever could have imagined in a foreign country. And these achievements were possible because I found a home away from home and family members to add to my biological family. I set out wanting to accomplish something and I now know that every accomplishment comes with hard work. My heroes were individuals who inspired me. These individuals were humble and led by example. The opportunity to be an educator and administrator in a foreign country allowed me to respect work and appreciate hard work. To do a good job, one must love it and acknowledge others who are an integral part of the accomplishment. It is also important to recognize the self. This is difficult because even when people think they have done a good job, they wait for others to acknowledge it in fear of seeming presumptuous, vain, or even arrogant. But self-congratulation is self-acknowledgement and is a recipe for continued good work. I learned this important lesson after retirement, when I had an opportunity to look back and take stock of what I had accomplished in my years away from home.

Receiving a prestigious University Professorship award in 2007 marked my entrance into the distinguished group of university professors who had been recognized before me for their contribution to the mission of the university and undergraduate education. Because this award is given to only one individual a year, I felt incredibly honored by the university for such a recognition. This group helped me to continue to grow in the profession and service to the University of Georgia in addition to allowing me to expand my circle of friends. I was surprised how easily and quickly I was integrated into this group of scholars compared to the Lilly Teaching Fellows program I was inducted into in 1990.

The Lilly Teaching Fellows Program was originally established at the University of Georgia in 1984 because of a grant from the Lilly Endowment. I was in the seventh group since its inauguration. The fellowship program was offered to faculty who were in their initial three years of tenure track at UGA. Its goal was to groom the best teachers and researchers who would enrich UGA's teaching mission. The program was extremely beneficial to young and upcoming faculty because it enabled them to meet regularly to discuss teaching and research agendas that significantly affected their tenure and promotion prospects. It also was an opportunity for faculty members to meet other assistant professors across campus. I attributed my inability to establish a professional connection within this group to the fact that all of us had a singular focus, how to survive the tenure chopping block at the end of six years of service at the University of Georgia. Unless one could find another institution to hire them, being denied tenure could be the end of the road to being accepted in academia. Thus, the Lilly Fellows program was not set up to create group synergy, and I felt like I had simply been incorporated in another university department. Despite the social activities, the individuals in this program were not necessarily interested in forging academic alliances but simply reconnecting with those they already knew in such a competitive environment. I also felt like every meeting or social event was less social, but rather a gathering of show and tell. The group dissipated at the end of our fellowship in 1991. Unfortunately, because I rarely attended reunions, I lost contact with the other participants in my group. By the time I retired in 2018, only four of us were still at UGA. Many left UGA before or after tenure to work in other institutions. As they left, one after the other, I felt that the intended outcome, which was largely encouraging faculty retention and progression was not realized by my group.

However, I gained some valuable experiences. What I treasured most was the leadership of the program's architect, Professor Ronald Simpson. His talks and advice about life at UGA were always inspiring. It allowed me to focus on how to discover and explore my abilities and strengths to be successful. I also cherish the advice he gave during one of his academic counseling sessions, saying that our colleagues in the department are not our friends. We should seek friends outside our departments and expose ourselves to new ideas and different people. It turned out to be the wisest advice I have ever received.

One person outside of my UGA circle was my primary-care physician. To navigate the confusing and all-consuming health care system in America warrants a doctor who cares for his patients and treats them as family. Dr. Farris T. Johnson Jr. is an important person in my journey. To have a personal physician is a privilege, something I did not have until I came to America. In Tanzania, if you want a personal doctor, you go to a private clinic or hospital, and even then, you can be viewed as a number, a contributor to the doctor's personal wealth. The patient-doctor relationship does not exist, except for those who can develop it through their fame or riding on the coattails of a well-known person. It is, unfortunately, a system of "who knows who." Most of the time spent in a doctor's office involves only physical evaluation and very little discussion. The most a patient might do is answer a couple of questions that the doctor asks, like, "What is the problem today?" or "When did the symptoms start?" After that, the doctor might do an exam or simply prescribe medication. It is refreshing in America to experience a doctor's attempt to make the patient proactive in their health care. It took me a long time to understand the system and to learn how to explain any symptoms I had. Dr. Johnson was not the first doctor I had in Athens; I had several before I arrived at his clinic. But Dr. Johnson found the cause of a persistent cough that had dogged me for ten years—Lisinopril, a medication I had been taking for ten years for high blood pressure. He told me to discontinue it well before it was officially recalled.

He was also my father's doctor when he came for a visit in 2004. It was Dr. Johnson's efforts that enabled my father to completely recover from a stroke that should have debilitated him and left him unable to retain his faculties or walk. My father always thought of Dr. Johnson as his angel who brought him back to life because he was not ready to die. Dr. Johnson is a man of faith, and it shows in his medical practice. He showed me the indomitable strength of the human body and mind and that life ends when one wants it to end.

The church I joined when I came to Athens became my community, separate from academia. I made a lot of friends like Dr. Jean Friedman, a history professor. Others include many women in the Social Graces Catholic group who were central to my campaign to support abandoned children through the One Kid at a Time (OKAT) Foundation. Many of their names are on the foundation's website, www.snyderupendo-okat.com. One special person in this group is Christine Reynolds who made a trip to Tanzania to visit the orphanage where these children are cared for. On her return, she established a circle of friends of OKAT and remained a dedicated fundraising organizer in support of activities that benefit the foundation.

To conclude, my life's journey has been nourished by my students, particularly those who participated in my study-abroad programs in Tanzania. The program, which started as a Maymester interdisciplinary program in Tanzania and then morphed into three programs—Maymester, Service Learning, and K-12 Teachers Group Project Abroad—created a large family of alumni. Participants consider the program integral to their respective scholarships. It has been more than twenty years since these programs started and many of the students say the experience was life changing. In a way, they remind me of my journey and what my time in America has brought to me. I am forever changed because of my time here. Many participants have remained in touch and continue to send updates about their lives. Some, like Maria-Harding Blanchard, decided to return to Tanzania for a year to teach at one of the partner institutions. Since then, she has joined forces in the foundation's mission to support displaced and disadvantaged children. She and her family are committed to this project and represent the pride and joy of her journey from America to Tanzania.

Equally valuable are all the people I have met along this journey; I could not name all of them here. Their care and friendship have been invaluable. I have built a circle of friends. By simply meeting one person, I gained several more incredible additions to my life. They were willing to introduce me to their family members too. As such, they allowed me to be in their social, not just academic, circles. In addition to my academic arena, I have made more friends around Athens, Georgia. I am grateful for the role each person has played and for making me a member of their village when my family was thousands of miles away. They did not allow me to be lonely. Each one has made me a better person than I could have ever dreamed. They have become my forever family members.

> People are lonely because they build walls instead of bridges.
> My life journey had many bridges built with numerous friends and well-wishers.

SAFARI NJEMA

I say to one, I say to all,
always try to build a bridge by reaching out,
connect with others.
It is worth the effort, and
it brings warmth,
friendship,
family,
and love
into your life.

Acknowledgements

My first and foremost gratitude is to my parents for giving me life. My father was very protective of my sister and me. He often reminded us that the world was not kind to women and that education was our salvation. He encouraged and supported us, and often told me the sky was the limit. A special thank you to my mother and all the female elders in my family. My success is their success. I escaped the nuances of the culture that looked down on them and became a person they did not expect. They had different expectations of me as a girl, especially as a first-born girl. My role in the family and other cultural expectations were not fully met as I left home at age eleven for boarding school to be educated by missionary nuns from England. I thank them for encouraging me rather than discouraging me from pursuing my ambitions.

It is impossible to name all the people who have touched my life in one way or another throughout this journey. My experience was enriched by their presence, both socially and professionally. I am cognizant of the fact that without their love, respect, and support, I would not have had such a rich experience. For those who know me, I hope you will agree that we are a product of our experiences. Special thanks go to Dr. Betty Jean Craige, Dr. Freda Scott-Giles, and Dr. Cherie Snyder. I am indebted to Dr. Craige who has been my friend and mentor for many years and who, in addition to encouraging me to write, penned the Forward, and connected me to the publisher. Other friends who enriched my life in America include my adoptive parents Charles Prael and his wife, Jean B. Prael. They welcomed and supported me through graduate school and made their home my home and their family my second family. Their unconditional love made it possible to survive the inescapable realities of being a foreign student in graduate school in America. Dr. Jean Friedman is another individual who helped me navigate life in the early days at the University of Georgia. My initial contacts with her were at the University of Georgia Catholic Center, and our friendship evolved into a mentor-mentee relationship. Other friends include Alexandra (Alex) Wright, Lisa Mathew and Todd Ferry, who made me appreciate being a teacher, Drs. Judy McWillie, Barbara McCaskill, and Johann Buis, whose life stories from South Africa made me rethink how to evaluate the privileges accorded to someone growing up in a free country, and Leslie Feracho. I cannot resist adding a comment about Dr. Feracho because of a life-changing trip I took with her and Dr. Giles to Cuba in 2018. Her fluency in Spanish made it possible for me to learn a lot about Cuba, a country that had much political influence on the policies created by my ancestral country of Tanzania in the 1970s. I was struck by the similarities in the social systems in Cuba, particularly the establishment of price-controlled government cooperative shops that rationed the distribution of food and commodities. What I saw in Cuba brought a realization of that country's influence on Tanzania and confirmed the strong friendships between Cuba's then-leader, Fidel Castro, and Tanzania's first president, Julius K. Nyerere.

When we visited rural Cuba, I saw more similarities in the agricultural sector to Tanzania's village experiment of the 1970s.

I would be remiss if I did not thank two significant women, Amsale Abegaz, my business manager at the Institute of African Studies for eight years (2000-08), and Gina Wein, the business manager at Stanford University. Abegaz made my work so much easier after struggling, single-handedly, for six years to establish and administer a sustainable African Studies program at UGA. Wein contributed to my soft landing at Stanford as I transitioned from being a graduate student to a professor. She made me feel at home from the day I arrived. For more than thirty years, I have remained a friend and a member of her family.

Also, I cannot thank Michele Taylor Sherwin enough for providing valuable edits and comments on the early drafts of this book, Judith McWillie, for editing the pictures, and Michael Hale for the cover picture and design.

Made in the USA
Columbia, SC
29 August 2024